Periodontal Disease

Frontiers of Oral Biology

Vol. 15

Series Editor

Paul Sharpe London

Periodontal Disease

Volume Editors

Denis F. Kinane Philadelphia, Pa.

Andrea Mombelli Geneva

26 figures, 16 in color, and 6 tables, 2012

KARGER Basel · Freiburg · Paris · London · New York · New Delhi · Bangkok · Beijing · Tokyo · Kuala Lumpur · Singapore · Sydney

Frontiers of Oral Biology

Denis F. Kinane
Department of Periodontics
School of Dental Medicine
University of Pennsylvania
240 South 40th Street
Philadelphia, PA 19104
USA

Andrea Mombelli
Division of Periodontology
University of Geneva
School of Dental Medicine
19 rue Barthélemy-Menn
CH-1205 Geneva
Switzerland

This book was generously supported by Colgate GABA Your partner in oral health

Library of Congress Cataloging-in-Publication Data

Periodontal disease / volume editors, Denis F. Kinane, Andrea Mombelli.
p. ; cm. -- (Frontiers of oral biology, ISSN 1420-2433 ; v. 15)
Includes bibliographical references and indexes.
ISBN 978-3-8055-9833-0 (hard cover : alk. paper) -- ISBN 978-3-8055-9834-7 (electronic version)
I. Kinane, Denis F. II. Mombelli, Andrea. III. Series: Frontiers of oral biology ; v. 15. 1420-2433
[DNLM: 1. Periodontal Diseases--microbiology. 2. Periodontal Diseases--therapy. 3. Anti-Infective Agents--therapeutic use. 4. Biofilms. 5. alpha-Defensins--immunology. W1 FR946GP v.15 2012 / WU 240]
LCclassification not assigned
617.6'32--dc23

2011032572

Bibliographic Indices. This publication is listed in bibliographic services, including Current Contents® and Index Medicus.

www.karger.com
Printed in Switzerland on acid-free and non-aging paper (ISO 9706) by Reinhardt Druck, Basel
ISSN 1420–2433
ISBN 978–3–8055–9833–0
e-ISBN 978–3–8055–9834–7

Contents

Foreword

Our understanding of the etiopathology of periodontal disease has changed greatly over the last decade through greater appreciation of the complexity found when examining both the host and pathogens. The species diversity within the microbial biofilm and the huge layered complexity of the innate, inflammatory and adaptive immune responses to this biofilm warrant study and discussion. The current volume comprises reviews from renowned experts in these fields, and aims to present a readable overview of this exciting and pertinent subject matter.

The volume opens with a review by Zijnge et al. of the subgingival biofilm structure, utilizing combined histology, confocal scanning fluorescent microscopy and fluorescent in situ hybridization to localize the most abundant species from different phyla and the species associated with periodontitis. The authors provide new insights into the structure and composition of subgingival biofilms and the nature of the extracellular matrix. They describe 'subgingival' biofilms produced in vitro that may be useful in future therapeutics testing and modeling of host-pathogen interactions.

Diaz then discusses the microbial diversity of the subgingival biofilm and describes a complex human subgingival environmental niche where microorganisms from the three domains of life meet to form diverse biofilm communities existing in close proximity to the host. Bacteria constitute the most abundant, diverse and ultimately well-studied component of these communities, with about 500 bacterial taxa reported to occur in this niche. Cultivation and molecular approaches continue to reveal the breadth and depth of subgingival biofilm diversity as part of an effort to understand the subgingival microbiome, the collection of microorganisms that inhabit the gingival crevices. This review presents a summary of current understanding of subgingival microbial diversity and an overview of experimental models used to dissect the functional characteristics of subgingival communities.

Innate cellular responses to the periodontal biofilm are critical to our understanding of the host response in periodontal disease, and Benakanakere and Kinane address this subject. In response to infection, the host's resources comprise the innate, inflammatory and adaptive immune systems, whose role is to provide the appropriate response to the offending microorganisms. In some cases, this will be little or no response – i.e. when 'commensals' are encountered – and in other cases it is a gradated response depending very much on the host's own determination of the pathogenic nature of the microbial insult: and herein lies the root of variation in host responses that govern individual susceptibility.

As discussed in depth in the articles of Diaz and Zijnge et al., gingival inflammation originates from responses to multiple microorganisms that comprise the biofilm rather than to just one highly virulent species. An understanding of the interaction of structural and defensive host cells with the biofilm is pivotal in understanding periodontal disease etiology and to developing tailored therapeutics. Thus, this chapter addresses the main structural cell exposed to the biofilm, i.e. epithelial cells, and the subsequent chapter by Scott and Krauss addresses the neutrophil response to biofilms. Neutrophils or polymorphonuclear leukocytes (PMN) are the most abundant leukocytes whose primary purpose as antimicrobial professional phagocytes is to kill extracellular pathogens. PMN and macrophages are phagocytic cell types which, along with other cells, effectively link the innate and adaptive arms of the immune response, and help promote inflammatory resolution and tissue healing. This review covers the innate and inflammatory functions of PMN and describes their importance in the integrity of the periodontium in health and disease.

Gorr addresses the role of antimicrobial proteins and peptides (AMP) in the host response to biofilm bacteria. AMP are an early component of the host-response produced by salivary glands, oral epithelial cells and neutrophils. Over 45 AMP have been identified in the oral cavity. All are found in saliva and several are also present in the gingival crevicular fluid. This review considers the differential regulation of AMP expression in periodontal disease and suggests that AMP panels may have a role in oral fluid diagnosis of periodontal disease and in monitoring treatment outcome.

Having addressed fundamental aspects of the etiopathogenesis, Apatzidou now considers modern approaches to non-surgical biofilm management. In contrast to microorganisms growing in a planktonic state, the inhabitants of a biofilm are effectively protected within this dense structure from host defense mechanisms and from therapeutic agents including antimicrobials. The mechanical removal of the microbial biofilm and the establishment of meticulous plaque control measures comprise the key elements for the success of non-surgical periodontal treatment. Apatzidou considers controversies and modern trends in non-surgical periodontal therapy, such as full-mouth approaches and power-driven instrumentation. As outcome variables, clinical, microbiological and immunological aspects are considered following different treatment protocols, in addition to cost-effectiveness.

Continuing in the therapeutic vein, Mombelli addresses advances in antibiotic use in the management of periodontal disease. Antibiotics are generally an efficient means of treating bacterial infections, and therefore an obvious candidate in the treatment of periodontal diseases. Systemically and locally administered antimicrobial agents of all kinds have been evaluated in multiple clinical trials. The vast majority of studies have tested antibiotics as adjuncts to non-surgical debridement. No regime has demonstrated superiority over systemically administered amoxicillin and metronidazole in the treatment of any clinically or microbiologically defined variant of periodontal disease. The author argues that in light of the excellent results of a combination therapy with well-established drugs, clinical trials should compare newly proposed protocols for periodontal therapy to a positive control

Traditional treatments have focused on repairing the damage induced by periodontal disease; however, clinicians have begun to utilize regenerative techniques to rebuild bone, cementum and the periodontal ligament. Kao and Fiorellini describe the potential of barrier membranes with bone grafts and pharmacologically based efforts to foster selective cell repopulation and the regrowth of osseous structures.

To close the therapeutic contributions and the volume itself, Hasturk et al. propose a paradigm shift in the pharmacological approach to periodontal disease management. They argue that while it is clear that the inflammatory response is a major determinant in susceptibility to periodontitis, our understanding of the relationship of the causal agents in periodontitis to the pathogenesis is not as clear – and therefore neither is the rationale for therapies based on this. The discovery of new families of lipid mediators of resolution of inflammation, the lipoxins and EPA- and DHA-derived chemical mediators, and the resolvins and protectins opens new avenues in the design of resolution-targeted therapies to control the unwanted side effects of excessive inflammation. The novel protective and therapeutic actions of pro-resolution lipid mediators following microbial challenge are mediated by regulation of the inflammatory response that has a direct impact on the organization of the biofilm (plaque) and could suggest a paradigm shift in clinical periodontal therapeutics.

All of these valuable and insightful contributions make this volume a timely update on the etiopathogenesis of periodontal disease and its relevance to modern periodontal therapeutics.

D.F. Kinane, Philadelphia, Pa.
A. Mombelli, Geneva

Kinane DF, Mombelli A (eds): Periodontal Disease. Front Oral Biol. Basel, Karger, 2012, vol 15, pp 1–16

Subgingival Biofilm Structure

Vincent Zijnge[a] · Thomas Ammann[b] · Thomas Thurnheer[b] · Rudolf Gmür[b]

[a]Department of Oral Microbiology, University of Umeå, Umeå, Sweden; [b]Institute of Oral Biology, Section of Oral Microbiology and Immunology, University of Zürich, Zürich, Switzerland

Abstract

Periodontitis is an inflammatory disease of the oral cavity initiated by a microbial biofilm (or 'dental plaque'). Subgingival biofilms in periodontal pockets are not easily analyzed without the loss of structural integrity. These subgingival plaques are structured communities of microorganisms with great phylogenetic diversity embedded in a self-produced extracellular polymeric matrix. For almost three decades, knowledge of the structure of plaque located below the gingival margin has been limited to landmark studies from the 1970s that were unaware of the breadth of microbial diversity we appreciate now. Only recently has technical progress – combining histology, confocal scanning fluorescent microscopy and fluorescent in situ hybridization to localize the most abundant species from different phyla and species associated with periodontitis – provided new insights into the architecture of subgingival biofilms. This review focuses on the structure and composition of subgingival biofilms and discusses current knowledge on the nature of the extracellular matrix. We describe further structural aspects of 'subgingival' biofilms produced in vitro that are gaining considerable interest as we search for models to investigate biofilm development, resistance to antibiotics, extracellular polymeric matrix composition and function, and reciprocal host-cell-to-biofilm interactions.

Microbial dental plaque, regardless of differences in location and composition, consists of adherent consortia of microorganisms (mostly bacteria, few archaea, viruses, yeasts, amoebae, and a virtually unknown, presumably large, population of bacteriophages) that fulfill the consensus definition of biofilms. This definition describes '. . .a bacterial biofilm as a structured community of bacterial cells enclosed in a self-produced (hydrated) polymeric matrix and adherent to an inert or living surface' [1]. Subgingival biofilms are 3D structured communities of bacteria that live attached to the surface of the root of teeth or dental implants, with their outer surface directly facing the gingival tissue. In a healthy periodontium, these sites are not accessible to bacteria. However, persistence of biofilm at the gingival margin and in the gingival sulcus leads to gingivitis, a reversible condition, which in susceptible patients may progress to periodontitis characterized by the irreversible loss of the tooth-supporting structures.

A key feature of this destructive inflammatory process is the formation of deep gingival pockets, which are colonized by the biofilm as they develop. The biofilm bacteria are embedded in a poorly understood extracellular matrix composed of exopolysaccharides, proteins and extracellular DNA. The formation of subgingival biofilms and its continuous adaptation to changing environmental conditions is governed by a dynamic equilibrium between the microorganisms, the cellular and humoral host defense, and a multitude of anabolic and catabolic products and signaling factors produced by both the microbiota and the periodontal tissues [2]. This results in a complex biofilm ecology where bacteria behave fundamentally differently than in suspension cultures where they exhibit a planktonic existence.

Max A. Listgarten, a pioneer of structural analyses of oral biofilms, while still at UPenn in 1994, published an excellent review on 'the structure of dental plaque' [2], which summarized the relevant work done up to that time point and highlighted the concepts on the relationship of biofilm structure to clinical status and the clinical relevance of biofilm composition and structure. The review was written at a time when new molecular microbiological techniques were beginning to conquer the dental field [3]. Eventually, this new development led to an explosion of information on the complexity and diversity of biofilm composition not imaginable at that time. It is the intent of this chapter to build on the review of Listgarten [2] and update our current understanding of subgingival biofilms in light of the progress that the technical and conceptual developments of the last 15 years have brought with regard to biofilm diversity, in vivo and in vitro biofilm architecture, and extracellular matrix composition.

Improved Analytical Procedures Reveal Complex Subgingival Biofilm Composition

Extensive culture analyses performed in the 1980s and early 1990s to determine the predominant cultivable subgingival microbiota showed that this biofilm may harbor as much as 10^9 bacteria and more than 100 different species in a single pocket [4, 5]. If the microbiota of entire study cohorts rather than individual pockets are considered, much larger diversity is apparent and suggests that, on a population basis, more than 500 bacterial species might be found in samples from the human oral cavity [4]. Some subgingival species, such as *Aggregatibacter actinomycetemcomitans, Porphyromonas gingivalis* and *Tannerella forsythia,* stood out in that they were found with increased prevalence and in clearly elevated numbers at diseased sites in comparison to healthy control sites. In 1996, these species were designated as periodontal pathogens in the Consensus Report of the World Workshop in Periodontology. Today they figure among the best-studied medically important bacteria, and in many respects have become model organisms. Nevertheless, it was clear all the time that other cultivable and non-cultivable subgingival species may be equally important and would merit further investigation [5]. Spirochetes, accounting for as much as 50% of the microscopically detectable subgingival microbiota, may serve as a prime example.

Do we need to describe the microbial community structure in a taxonomically precise and quantitative way? Amann and Ludwig [6] raised this question in a review 10 years ago and answered it with: 'Yes, of course!' Why? Because complex microbial ecologies cannot be characterized sufficiently by selected individual taxa that may have adapted particularly well to their ecological niche. Complex microbial ecologies are rather defined by all the mutual interactions that determine the abundance, localization and activities of their members [6]. Clearly, new tools for studies of (oral) microbial diversity were required. These became available in the mid-1980s with the emergence of new procedures for the comparative sequencing of homologous biopolymers [7]. In particular the extensive sequencing of ribosomal RNA (16S and to a lesser extent 23S rRNA) became very attractive and influential as mirrored by the now more than 1.9×10^6 deposited 16S rRNA sequences (Ribosomal Database Project August 2011; http://rdp.cme.msu.edu). In 2001, Paster et al. [8] described 215 novel phylotypes associated with subgingival plaque. In the following years, numerous conceptually and technically similar studies complemented each other and lengthened the list of newly recognized species and phylotypes, culminating in the creation and most recently in an update of the Human Oral Microbiome database (www.homd.org) [9]. This database includes a backbone library of about 800 16S rRNA gene sequences that are structured phylogenetically and taxonomically into 619 taxa in 13 phyla. The backbone library will be expanded in the near future by data from an additional 36,000 clonal sequences of which approximately 10% – accounting for as much as 654 new taxa – are not yet covered by the current Human Oral Microbiome backbone database [9]. In addition to this, recent progress in DNA sequencing technology ('pyrosequencing') has brought an enormous increase in nucleotide sequences of oral microbial origin, resulting in an overwhelming number of phylotypes of unknown species affiliation [10–12]. Where in the biofilms are the taxa located? How abundant are they? What are their morphologies, their physiological activities, their importance for both biofilm and host-biofilm relationship? To a large part we do not know – yet. Recently, technical progress has provided very promising possibilities for the single-cell identification and localization in intact natural subgingival biofilms of most taxa and phylogenetic clusters with an available 16S rRNA sequence.

Methods to Study Subgingival Biofilm Architecture

Some of the most valuable information on supragingival biofilm formation, ecology and architecture has come from in situ studies performed with volunteers wearing expoxy crowns [13], artificial surfaces attached to selected teeth [14], or specifically designed appliances into which enamel, dentin, glass or plastic slabs had been inserted [e.g. 15–17]. While all these studies could only identify the biofilm bacteria based on morphology or cell wall structure (if electron microscopy was used) and differentiate between live and dead organisms using vitality stains, some more recent studies used fluorescent in situ hybridization (FISH) technologies to identify targeted biofilm bacteria at the species level [e.g. 18–21]. The application of phylogenetic group- or species-specific single cell identification techniques to undisturbed biofilms formed supragingivally in the oral cavity on retrievable surfaces is currently the state of the art, and promises to reveal important new information on supragingival plaque formation and architecture.

Unfortunately, the study of subgingival plaque is much more difficult in comparison to supragingival plaque due the protected location of periodontal pockets. Accordingly, the development of the 3D subgingival biofilm structure is less well characterized. Access to undisturbed natural subgingival biofilms can only be gained by tooth extraction. Listgarten and co-workers [22] pioneered structural analyses of oral biofilms more than three decades ago using light and electron microscopy. The major hindrance of their groundbreaking studies was the impossibility at the time to identify the detected bacteria beyond cell morphology (cell shape, cell wall structure and Gram stain). This obstacle has only recently been overcome when entire teeth affected by advanced periodontitis were extracted without disturbance of the adherent subgingival biofilm and – after immediate fixation and processing to serial sections of 2 μm thickness – were stained by FISH using various combinations of group- and species-specific rRNA probes and studied by epifluorescence or confocal laser scanning microscopy (CLSM) [23]. A way to investigate in situ subgingival biofilm formation without the loss of the tooth under study was developed by Wecke et al. [24], and over the last few years has been refined and applied in several studies [25–27]. This procedure uses gold foil or small plastic carriers covered with expanded polytetrafluoroethylene membranes, which are inserted to the depth of the periodontal pockets for defined periods of time and then processed similarly to extracted teeth. In the following, these recently published studies are reviewed in greater detail as they may lead the way to further investigations that likely will be carried out using much broader selections of phylogenetic group-, genus- and species-specific FISH probes to elucidate the subgingival biofilm architecture.

Subgingival Plaque Structure

A profound understanding of the in vivo structure of subgingival biofilms, i.e. the natural situation, is essential for designing and interpreting diversity measurements and in vitro experimental biofilms. Histological slices perpendicular to the root surface reveal three easily envisioned and clinical relevant areas, namely the bottom of the pocket, the root surface and the epithelial side [28]. At the bottom of the pocket, the periodontal ligament and the gingival connective tissue border the biofilm. This is

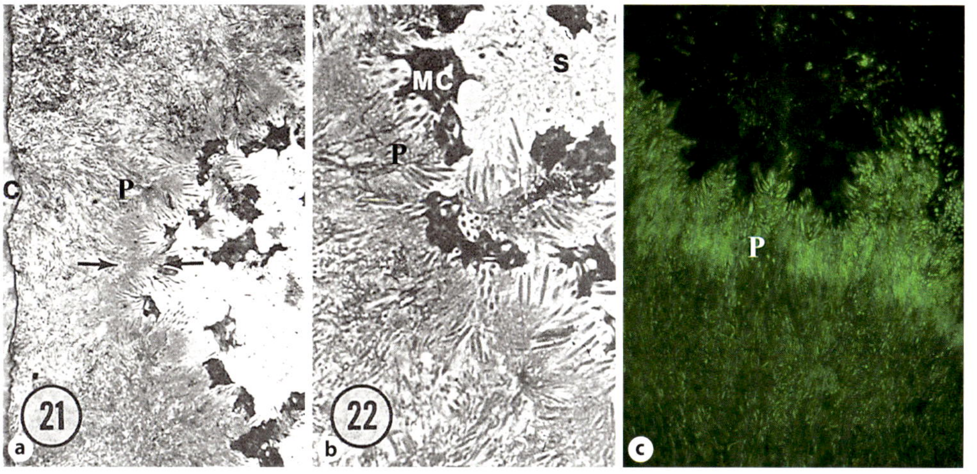

Fig. 1. Overview of subgingival biofilms taken in the 1970s and 2008 (**c**) from extracted teeth. **a**, **b** Electron microscopic views. Reproduced with permission of the American Academy of Periodontology [21, 22]. **c** Stained with a FITC-labeled universal bacterial probe that stains the bacteria green. Note the remarkable agreement in biofilm structure and observed cell morphologies. C = Cementum surface; MC = mammalian cells; P = palisading bacteria; S = spirochete-rich region.

the area where further attachment loss occurs in progressing periodontitis and where further pristine tooth surface becomes colonized. The species or bacterial cell morphologies observed in this area are filamentous, large rod-shaped, spirochete-shaped and branching. Especially the branching cells are abundant, whereas spirochetes are sometimes missing [29].

The bulk of the biofilm mass is situated along the root surface with histologically defined parts adhering to the root surface and facing the gingival host tissue. Listgarten [22] was the first to describe the structure of such subgingival biofilms in 1976 using light and electron microscopy. His pioneering work with natural teeth affected by periodontal disease provided the first informative glimpses at subgingival plaque structure. The pictures of subgingival biofilms showed a wealth of different cell morphologies like cocci, rods, fusiforms, spirochetes, flagellated bacteria, small and very large forms, and bacterial aggregates. Moreover, the biofilm itself showed a distinct organization, indicating that biofilms might be structured entities themselves that are beneficial to their inhabitants.

The images of figure 1 show a remarkable and unprecedented agreement between samples taken 30 years apart and analyzed with different techniques [22, 23]. They provide a clear overview of the different structures that can be distinguished when analyzing subgingival biofilms. Three different zones are observed between the root surface and the gingival tissue lining of the periodontal pocket. In the zone directly adhering to the root surface, relatively small cells without any particular orientation are visible (fig. 1, 2). These bacteria are embedded in somewhat fibrous membrane-

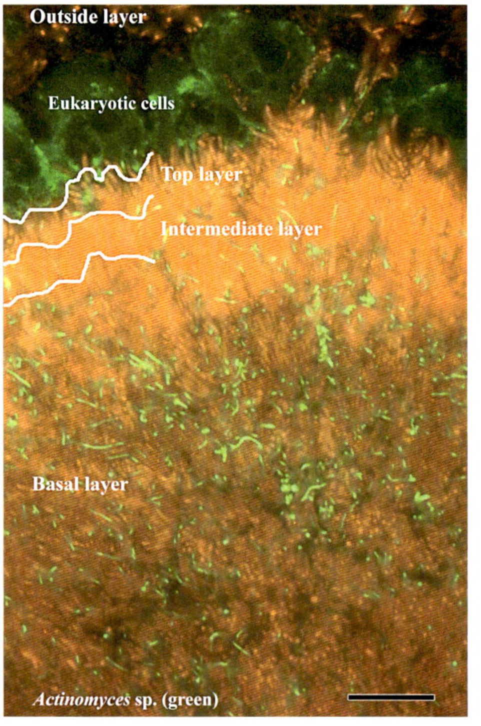

Fig. 2. Overview of the subgingival biofilm with *Actinomyces* sp. (green), bacteria (red) and eukaryotic cells (large green cells on top). The four different layers of the biofilm are depicted with the root surface orientated to the bottom of the image [for technical details, see 23]. Scale bar = 10 μm.

like material that may consist of debris, dead cells and biofilm matrix components. Amongst these unidentified bacteria with various morphologies, other bacteria can be recognized as *Actinobacteria* due to their characteristic branching or labeling by FISH (fig. 2) [23]. On top of that layer, a structurally different layer of bacteria can be observed that is more compact and electron dense (fig. 1, 2). This typical intermediate layer contains distinctive bacteria that are filamentous or rod shaped, and is without a well-defined intercellular matrix [22, 30]. FISH has identified these bacteria in part as *Fusobacterium nucleatum* and *T. forsythia* (fig. 3) [23]. The outermost part of the adhering biofilm that is facing the host tissues is characterized by a palisading layer of so-called 'test tube brushes' and of large parallel cells that are orientated with their long axis perpendicular to the root surface (fig. 1, 2). The test tube brushes consist of one or several large central filaments surrounded by gram-negative rods or small filaments. The parallel-orientated cells have been identified as members of the *Synergistetes* phylum (fig. 4) [23, 31], and represent a lineage composed of taxa that for the most part are not yet cultivable. Opposing the *Synergistetes* in this outermost part of the adhering biofilm, polymorphonuclear neutrophils can be observed by light, electron and epifluorescence microscopy (fig. 1, 2, 4, 5) [22, 28]. Several reports described an additional loose layer of bacteria covering the adhering biofilm without a clear cellular organization and dominated

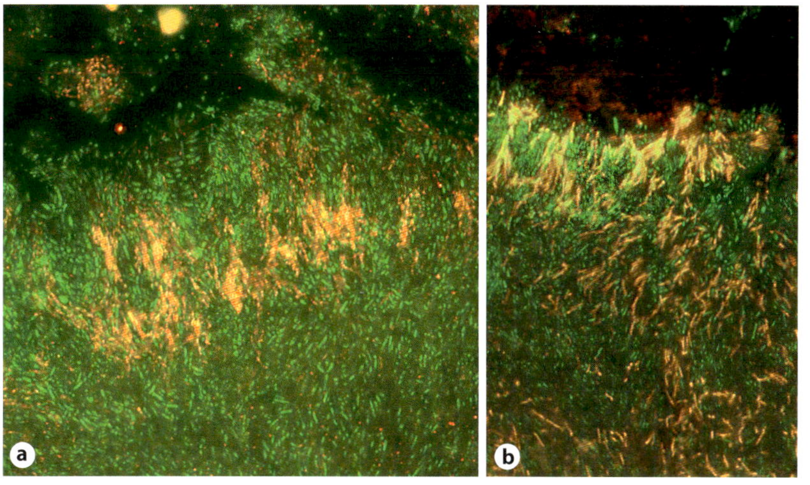

Fig. 3. The intermediate layer of the biofilm is dominated by fusobacteria (yellow, **a**) and *T. forsythia* (yellow, **b**). Modified from Zijnge et al. [23]. Scale bars = 10 μm.

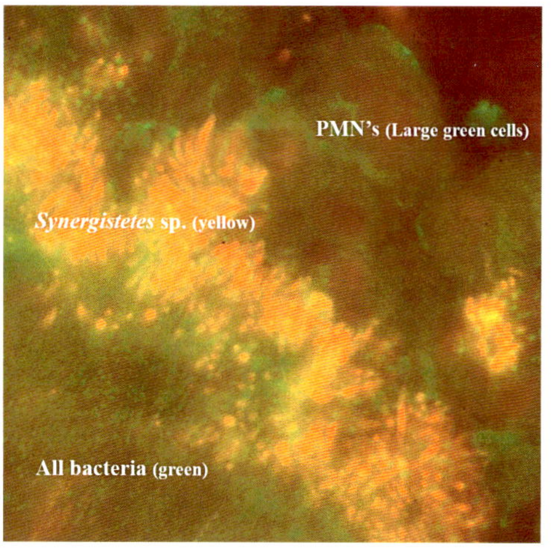

Fig. 4. Detailed overview of the top layer of the biofilm where *Synergistetes* sp. (yellow), visible as large perpendicularly orientated cells, are opposing large eukaryotic cells (green), most likely polymorphonuclear leukocytes. Modified from Zijnge et al. [23]. Scale bars = 10 μm.

by flagellated and spirochetal cells belonging to *Selenomonas* and *Treponema* species [23, 26, 32], although rods, filaments, cocci and test tube brushes could also be observed (fig. 6). A clear demarcation between the host tissue and this loose bacterial layer is not recognizable, probably because severe inflammation induces immunopathological tissue damage accompanied by collagen loss, ulceration and marked tissue permeabilization.

The subgingival biofilm shows a stratification from the root surface towards the epithelium lining. The basal, intermediate, top and outer layers seem to provide a

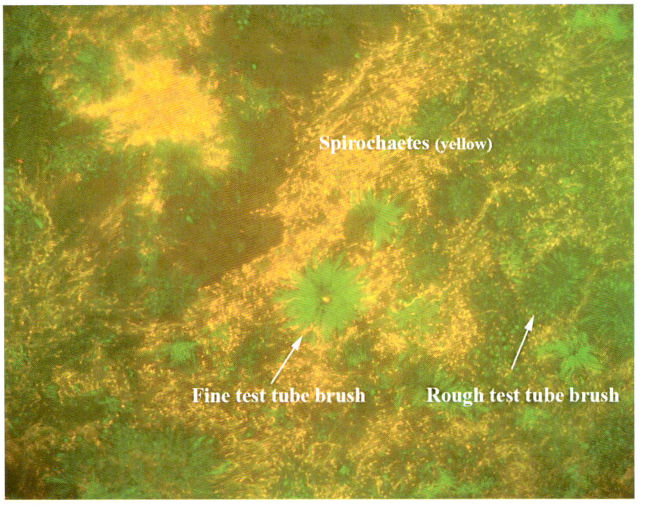

Fig. 5. Detail of the outside layer showing the dominance of *Treponema* sp. (yellow) and the presence of two types of test tube brush (arrows). Modified from Zijnge et al. [23].

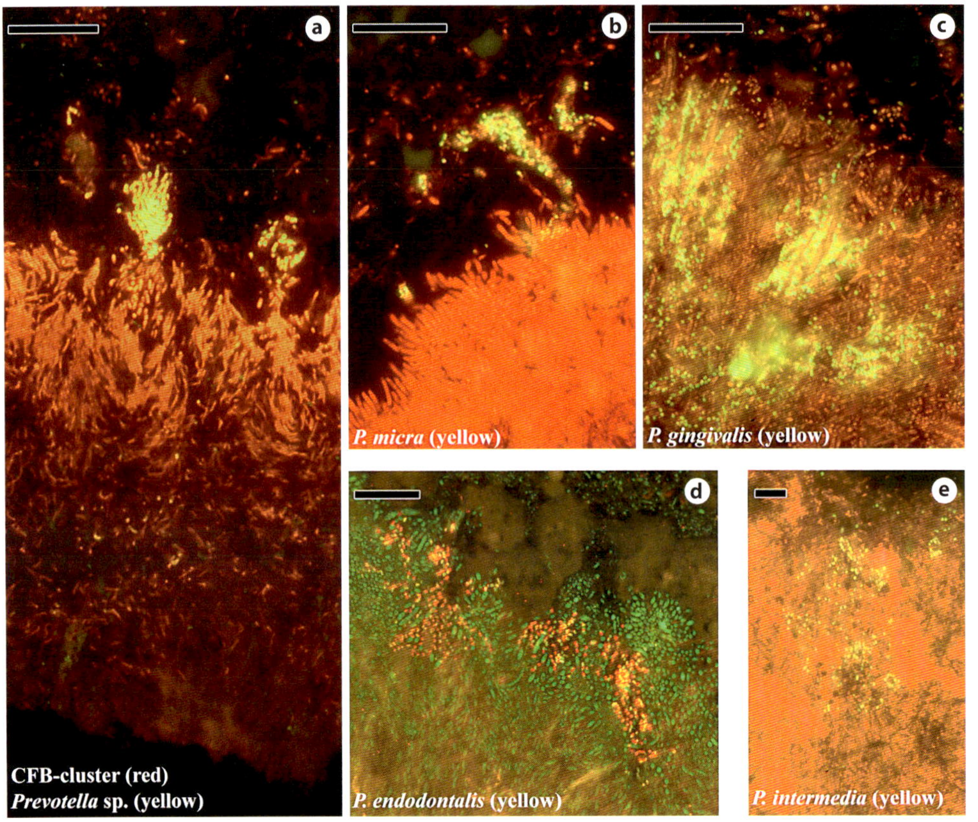

Fig. 6. Localization of typical periodontitis associated species in the biofilm. **a** Overview of the subgingival biofilm with *Cytophaga-Flavobacterium-Bacteroides* (CFB)-cluster species (red) and *Prevotella* sp. (yellow). Since *Prevotella* sp. are part of the CFB-cluster of bacteria, cells appear in yellow. **b** Top of the biofilm with a microcolony of *P. micra* (yellow). **c** Microcolonies of *P. gingivalis* (yellow) in the top layer. **d** Microcolonies of *P. endodontalis* (yellow) in the top layer. **e** Microcolonies of *P. intermedia* in the top layer. In panels **b–e**, bacteria are universally stained with a red or green label. Reproduced from ZIJnge et al. [23].

general structure of the biofilm with corresponding species associated with each layer. *Actinomyces* sp., *F. nucleatum*, *T. forsythia*, *Synergistetes* sp. and *Treponema* sp. appear to make up a framework of the subgingival biofilm. Observations with species-specific probes and gold-labeled antibodies further show species that colonize within this biofilm framework by forming distinct microcolonies. In the outer layer of the adhering biofilm, closest to the epithelium and cells of the host defense, aggregates or microcolonies of *Prevotella intermedia*, *Porphyromonas endodontalis*, *P. gingivalis* and *Parvimonas micra* (formerly *Peptostreptococcus micros*) have been observed (fig. 5).

Microbial biofilms are dynamic communities exposed to an ever-changing environment. Therefore, the outlined architecture of the subgingival biofilm should not be seen as a rigid structure, but rather as a continuously changing consortium that is influenced by bacterial growth, attachment and detachment, the host's inflammatory immune response, the host's oral hygiene measures, and the nutritional conditions defined by the ecological niche. The availability or lack of nutrients is of course directly linked to both biofilm composition and extent of inflammation. Although the general setup of the subgingival biofilm framework appears relatively uniform, less common species may be observed in diverse sites, in particular within the loose layer that covers the adhering biofilm.

Clearly, our understanding of the subgingival plaque structure is incomplete. The number of teeth studied, the number of patients who provided teeth, and the number and specificities of the employed FISH probes are too limited for more precise conclusions. The degree of diversity currently appreciated among plaque organisms [9] is not reflected in the currently available studies of biofilm architecture. The application of rRNA FISH probes with specificities for large phylogenetic groups – such as β-, γ- or ε-Proteobacteria, *Bacteroidetes*, *Eubacteria* sp., *Selenomonas* sp. – to subgingival plaque samples that had been dispersed and subjected to FISH on multi-well glass slides, indicates that other bacterial taxa may account for substantial amounts of the biofilm [Gmür et al., unpubl. data].

Structure of 'Subgingival' Biofilms Formed in vitro

It is evident that these natural subgingival biofilms are extremely complex and sited in a barely accessible, fluctuating and inflammation-affected environment. Together with ethical restrictions, this limits in vivo experimentation and makes interpretation of results difficult. Several research groups have developed in vitro model systems of subgingival plaque with the aim of reducing complexity while maintaining the bacterial biofilm 'lifestyle' and reproducing characteristic properties of such communities [e.g. 33–36]. Regardless whether freshly collected dispersed human subgingival plaque or balanced mixtures of strains of subgingival species were used to initiate biofilm formation, the nutritional conditions were recognized to have a dominant influence on

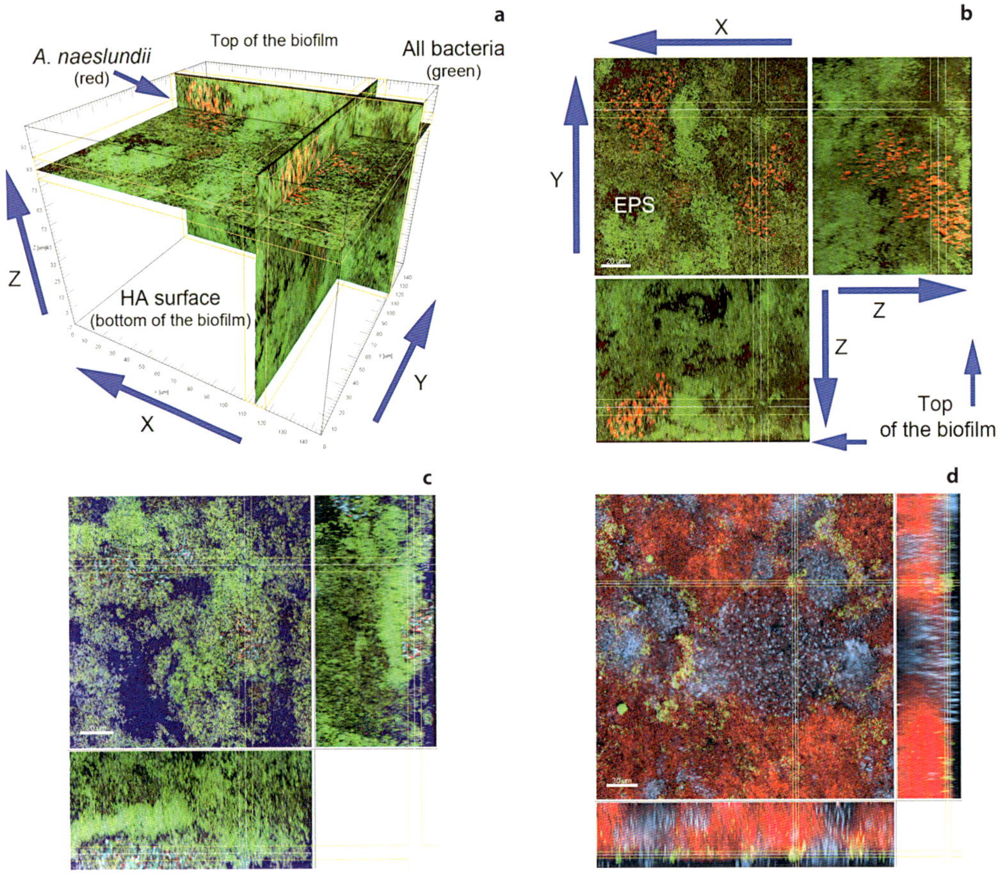

Fig. 7. CLSM images of 10-species 'subgingival' in vitro biofilms grown in parallel for 64.5 h on hydroxyapatite (HA) discs in 50% serum/50% medium and then fluorescence-labeled by multiplex FISH. **a** 3D arrangement of transverse (X-Y), sagittal (Y-Z) and coronal (X-Z) CLSM sections shown in the other panels. **a**, **b** Sections through the same biofilm stained for *A. naeslundii* (red, probe L-Act476–2-Cy3) and all bacteria (green, SYTOX green/YO-PRO). Black areas are bacteria-free spaces filled by extracellular polymeric substances (EPS). Scale bar = 20 µm. **c** Biofilm stained for *T. denticola* (light blue, TrepG1_679-Cy5), *P. gingivalis* (red, L-Pgin1006–2-C3), all bacteria (green, Sytox and YO-PRO) and extracellular polysaccharides (dark blue, Calcofluor). Note that the two specifically labeled species are located in close proximity to each other near the surface of the biofilm. Scale bar = 20 µm. **d** Spatial distribution of *S. anginosus and S. oralis* (both blue, STR405-Cy5) and *F. nucleatum* (red, Fnuc133c-Cy3) demonstrating their abundance and formation of large self-aggregates. The green-yellow stained bacteria are *V. dispar* (VEI217-FAM), characteristically located in the neighborhood of streptococci and at the top of the biofilm. Scale bar = 30 µm. Images are from Ammann et al. [unpubl. data].

biofilm development [35, 36]. This must have profound effects on biofilm structure, but information on the architecture of such in-vitro-generated subgingival biofilms is still scarce [34, 36]. Figure 7 shows images from our ongoing studies [Ammann et al., unpubl. data] investigating the architecture of 10-species 'subgingival' biofilms grown in vitro in a saliva-free standard fluid medium supplemented with human serum (50:50%). The biofilms reached a thickness of 70–100 μm after 64.5 h of incubation with all inoculated species at detectable levels. *F. nucleatum, Streptococcus oralis* and *Streptococcus anginosus* predominated and were found throughout the biofilm (fig. 7d). Streptococci did not grow dispersed but formed large compact aggregates, often surrounded by layers of *Veillonella dispar* (fig. 7d). *A. naeslundii* was detected in relatively large individual colonies (fig. 7a, b) embedded in a readily detectable extracellular polysaccharide matrix (not shown) and spreading from the bottom third to the top of the biofilm. This suggests that *A. naeslundii* expanded rapidly from a relatively limited number of foci and, with a little longer incubation period, would have possibly spread over the entire body of the biofilm. This would correspond to the spatial distribution seen with in vivo biofilms (fig. 2). In contrast, *Treponema denticola* and *P. gingivalis* were detected selectively in close proximity to each other atop of very dense brightly stained cushions of bacteria (presumably streptococci) at the biofilm surface (fig. 7C). The presence of spirochetes and *Porphyromonas* species at this location is a characteristic attribute of subgingival plaque (fig. 5, 6); hence, it will be important to further analyze in vitro the reproducibility of this interesting finding. It must be emphasized that these are preliminary data gained with a limited number of biofilms. However, they demonstrate that the combination of 3D CLSM and multiplex FISH promises to be of great value in further elucidating the architecture of both natural and artificially generated subgingival biofilms.

Extracellular Matrix Composition

There are as many different types of biofilms as there are bacteria [37]. Nevertheless, in any biofilm a substantial part consists of material other than bacteria. In general it is estimated that microorganisms account for less than 10% of the dry weight of biofilms and that the extracellular matrix contributes over 90% [38]. The formation of the extracellular matrix is generally recognized as the second stage during biofilm formation, after initial adherence and proliferation of microorganisms [39]. The extracellular matrix is composed of a heterogeneous mixture of polysaccharides, proteins and extracellular DNA (eDNA), which are called collectively extracellular polymeric substances (EPS). These complex conglomerate EPS are very difficult to analyze. Hence, it is not surprising that current knowledge of biofilm-associated EPS is derived almost exclusively from single-species biofilm models. However, whether the EPS production by these bacteria is the same when they live within a complex 'biofilm-city' [40] remains to be seen.

Extracellular polysaccharides occur in two basic forms. They can be associated with the cell surface and form a capsule or be secreted as a slimy biofilm matrix. Most polysaccharides are long linear or branched molecules composed of multiple saccharide units like glucose, fructose or sucrose and exist as homo- or heteropolysaccharides.

Many of them possess acyl-groups and organic and inorganic substituents like acetate, pyruvate or sulphate. These substituents and the presence of charged sugar residues largely determine the physical properties of the extracellular polysaccharides. The type of polysaccharide that is produced usually varies among the different species and some species can produce multiple kinds of polysaccharides [37]. One of the most commonly studied matrix polysaccharides is β-1,6-*N*-acetyl-D-glucosamine called PNAG or PGA. It is produced, among others, by *A. actinomycetemcomitans* for biofilm formation. In *A. actinomycetemcomitans* PGA mediates intercellular adhesion and contributes to biofilm cohesion, but also protects the cells against killing by macrophages [41]. *Prevotella nigrescens* produces a heteropolysaccharide that is composed mainly of mannose, but also contains other sugars, including glucose and fructose [42]. A mannose-rich polysaccharide has also been identified from *P. intermedia* [43, 44]. The polysaccharides produced by *P. nigrescens* and *P. intermedia* contribute to biofilm formation and the resistance to neutrophil phagocytosis. *P. gingivalis* strains on the other hand produce a capsular polysaccharide. Encapsulated strains have been shown to evade the immune system [45] but non-encapsulated strains are more adherent to epithelial cells and show strong autoaggregation and enhanced biofilm formation [46, 47]. If and what type of free extracellular polysaccharide *P. gingivalis* produces for the formation of biofilms is unknown. Actinobacteria produce levan which is a homopolysaccharide composed of β(2r6)-linked fructosyl units [48]. Levan may function as extracellular storage polymers, but has also been shown to stimulate the inflammatory response. The contribution of levans to actinobacteria biofilm formation is unknown.

Extracellular proteins that can be found in, and contribute to, a biofilm are lectins and sugar-binding proteins that facilitate cell-to-cell or cell-to-matrix interactions [49]. A second group of extracellular proteins involved in cell-to cell or cell-to-matrix interactions are autotransporters. Autotransporters transport themselves across the outer membrane of gram-negative bacteria and may function in adherence and biofilm formation. A third group of extracellular proteins that structurally contribute to the biofilm matrix are pili, also called fimbriae [37]. A type I and a type IV pilus are distinguished. The type I pilus consists of a long rigid structure of repeating subunits that is attached to the cell with a fimbrial tip that recognizes specific substrates. Type I pili have been observed on a number of oral bacteria [50] including *P. gingivalis*, *A. naeslundii* [51] and *P. intermedia* [52]. In *Actinomyces oris*, type I pili are functional in biofilm formation [53]. Type IV pili are often located at one of the poles of the cell and may provide the cell with twitching motility. The pilus is a fiber composed of repeating units of pilin encoded by the *PilA* gene [54]. Type IV pili have been identified in *Eikenella corrodens* and *A. actinomycetemcomitans* [50, 55]. Genomic screening indicated the presence of type IV pili as in *F. nucleatum* [54]. In *A. actinomycetemcomitans*, the type IV pilus is considered to be a distinct subtype assembled as bundles of long thin fibrils encoded by the *tad* locus [55]. *A. actinomycetemcomitans* strains deficient in the formation of pili form relatively fragile biofilms [56].

Extracellular DNA has been shown to be an important constituent of biofilms and to contribute to biofilm integrity [38, 57, 58]. There are indications that eDNA results from the controlled lysis of cells as well as from the active release of DNA containing membrane vesicles by viable cells [37, 38]. It is tempting to speculate that eDNA might function as a grid along which bacteria can move through the biofilm by using type IV pili with DNA-specific binding sites. There is no doubt that the characteristics of the EPS contribute to the specific properties of the biofilm. Many components are hygroscopic and keep the biofilm hydrated and prevent the cells from drying. The EPS matrix can act as a molecular sieve and protect the bacteria in the biofilm from toxic compounds or host immune responses, and nutrients are accumulated. The enhanced resistance of biofilms in comparison to planktonic bacteria to antimicrobial agents is well established [59]. The entanglement of biopolymers provides the biofilm with the physical strength and flexibility to withstand mechanical forces and shear stress. While biofilms provide an optimal environment for bacterial growth and survival, external impacts that may cause starvation or altered bacterial cross-talk are known to induce the production of enzymes that can potentially degrade EPS components. For example, *A. actinomycetemcomitans* produces the extracellular *N*-acetyl-β-hexoseaminidase dispersin B that hydrolyzes the glycosidic links in PNAG and degrades the biofilm matrix. This may result in the dispersal of cells or clumps of cells, which in turn can colonize new surfaces and initiate a new biofilm lifecycle [38, 57]. Flemming and Wingender [38] concluded in their recent review that 'despite much research on biofilms, basic questions remain' and continued that 'a better understanding of the regulation of EPS production in mixed-species biofilms, as well as a spatial and temporal dissection of the phases in EPS production, will reveal important aspects of the oldest, most successful and widespread form of life on Earth'.

Acknowledgement

T.A. is supported by the research fund of the Swiss Dental Association.

References

1 Costerton JW, Stewart PS, Greenberg EP: Bacterial biofilms: a common cause of persistent infections. Science 1999;284:1318–1322.

2 Listgarten MA: The structure of dental plaque. Periodontol 2000 1994;5:52–65.

3 Tanner A, Maiden MFJ, Paster BJ, Dewhirst FE: The impact of 16S ribosomal RNA-based phylogeny on the taxonomy of oral bacteria. Periodontol 2000 1994;5:26–51.

4 Moore WEC, Moore LVH: The bacteria of periodontal diseases. Periodontol 2000 1994;5:66–77.

5 Haffajee AD, Socransky SS: Micobial etiological agents of destructive periodontal diseases. Periodontol 2000 1994;5:78–111.

6 Amann R, Ludwig W: Ribosomal RNA-targeted nucleic acid probes for studies in microbial ecology. FEMS Microbiol Rev 2000;24:555–565.

7 Amann RI, Binder BJ, Olson RJ, Chisholm SW, Devereux R, Stahl DA: Combination of 16S rRNA-targeted oligonucleotide probes with flow cytometry for analyzing mixed microbial populations. Appl Environ Microbiol 1990;56:1919–1925.
8 Paster BJ, Boches SK, Galvin JL, Ericson RE, Lau CN, Levanos VA, Sahasrabudhe A, Dewhirst FE: Bacterial diversity in human subgingival plaque. J Bacteriol 2001;183:3770–3783.
9 Dewhirst FE, Chen T, Izard J, Paster BJ, Tanner AC, Yu WH, Lakshmanan A, Wade WG: The human oral microbiome. J Bacteriol 2010;192:5002–5017.
10 Keijser BJ, Zaura E, Huse SM, van der Vossen JM, Schuren FH, Montijn RC, ten Cate JM, Crielaard W: Pyrosequencing analysis of the oral microflora of healthy adults. J Dent Res 2008;87:1016–1020.
11 Zaura E, Keijser BJF, Huse SM, Crielaard W: Defining the healthy 'core microbiome' of oral microbial communities. BMC Microbiol 2009;9:259.
12 Contreras M, Costello EK, Hidalgo G, Magris M, Knight R, Dominguez-Bello MG: The bacterial microbiota in the oral mucosa of rural Amerindians. Microbiology 2010;156:3282–3287.
13 Listgarten M, Mayo HE, Tremblay R: Development of dental plaque on epoxy resin crowns in man. J Periodontol 1975;46:10–26.
14 Wood SR, Kirkham J, Marsh PD, Shore RC, Nattress B, Robinson C: Architecture of intact natural human plaque biofilms studied by confocal laser scanning microscopy. J Dent Res 2000;79:21–27.
15 Netuschil L, Reich E, Unteregger G, Sculean A, Brecx M: A pilot study of confocal laser scanning microscopy for the assessment of undisturbed dental plaque vitality and topography. Arch Oral Biol 1998;43:277–285.
16 Zaura-Arite E, van Marle J, ten Cate JM: Confocal microscopy study of undisturbed and chlorhexidine-treated dental biofilm. J Dent Res 2001;80:1436–1440.
17 Wood SR, Kirkham J, Shore RC, Brookes SJ, Robinson C: Changes in the structure and density of oral plaque biofilms with increasing plaque age. FEMS Microbiol Ecol 2002;39:239–244.
18 Foster JS, Palmer RJJ, Kolenbrander PE: Human oral cavity as a model for the study of genome-genome interactions. Biol Bull 2003;204:200–204.
19 Palmer RJ Jr, Diaz PI, Kolenbrander PE: Rapid succession within the *Veillonella* population of a developing human oral biofilm in situ. J Bacteriol 2006;188:4117–4124.
20 Al-Ahmad A, Wunder A, Auschill TM, Follo M, Braun G, Hellwig E, Arweiler NB: The in vivo dynamics of *Streptococcus* spp., *Actinomyces naeslundii, Fusobacterium nucleatum* and *Veillonella* spp. in dental plaque biofilm as analysed by five-colour multiplex fluorescence in situ hybridization. J Med Microbiol 2007;56:681–687.
21 Dige I, Raarup MK, Nyengaard JR, Kilian M, Nyvad B: *Actinomyces naeslundii* in initial dental biofilm formation. Microbiology 2009;155:2116–2126.
22 Listgarten MA: Structure of the microbial flora associated with periodontal health and disease in man: a light and electron microscopic study. J Periodontol 1976;47:1–18.
23 Zijnge V, van Leeuwen MBM, Degener JE, Abbas F, Thurnheer T, Gmür R, Harmsen HJM: Oral biofilm architecture on natural teeth. PLoS ONE 2010;5:e9321.
24 Wecke J, Kersten T, Madela K, Moter A, Göbel UB, Friedmann A, Bernimoulin J-P: A novel technique for monitoring the development of bacterial biofilms in human periodontal pockets. FEMS Microbiol Lett 2000;191:95–101.
25 Moter A, Göbel UB: Fluorescence in situ hybridization (FISH) for direct visualization of microorganisms. J Microbiol Methods 2000;41:85–112.
26 Drescher J, Schlafer S, Schaudinn C, Riep B, Neumann K, Friedmann A, Petrich A, Göbel UB, Moter A: Molecular epidemiology and spatial distribution of *Selenomonas* spp. in subgingival biofilms. Eur J Oral Sci 2010;118:466–474.
27 Schlafer S, Riep B, Griffen AL, Petrich A, Hübner J, Berning M, Friedmann A, Göbel UB, Moter A: *Filifactor alocis* – involvement in periodontal biofilms. BMC Microbiol 2010;10:66.
28 Waerhaug J: Subgingival plaque and loss of attachment in periodontosis as observed in autopsy material. J Periodontol 1976;47:636–642.
29 Saglie R: A scanning electron microscopic study of the relationship between the most apically located subgingival plaque and the epithelial attachment. J Periodontol 1977;48:105–115.
30 Westergaard J, Frandsen A, Slots J: Ultrastructure of the subgingival microflora in juvenile periodontitis. Scand J Dent Res 1978;86:421–429.
31 Vartoukian SR, Palmer RM, Wade WG: Diversity and morphology of members of the phylum 'Synergistetes' in periodontal health and disease. Appl Environ Microbiol 2009;75:3777–3786.
32 Noiri Y, Li L, Ebisu S: The localization of periodontal-disease-associated bacteria in human periodontal pockets. J Dent Res 2001;80:1930–1934.

33 Bradshaw DJ, Marsh PD, Watson GK, Allison C: Role of *Fusobacterium nucleatum* and coaggregation in anaerobe survival in planctonic and biofilm oral microbial communities during aeration. Infect Immun 1998;66:4729–4732.

34 Hope CK, Wilson M: Biofilm structure and cell vitality in a laboratory model of subgingival plaque. J Microbiol Methods 2006;66:390–398.

35 Walker C, Sedlacek MJ: An in vitro biofilm model of subgingival plaque. Oral Microbiol Immunol 2007;22:152–161.

36 Guggenheim B, Gmür R, Galicia JC, Stathopoulou P, Benakanakere MR, Meier A, Thurnheer T, Kinane D: In vitro modeling of host-parasite interactions: the 'subgingival' biofilm challenge of primary human epithelial cells. BMC Microbiol 2009;9:280.

37 Karatan E, Watnick P: Signals, regulatory networks, and materials that build and break bacterial biofilms. Microbiol Mol Biol Rev 2009;73:310–347.

38 Flemming HC, Wingender J: The biofilm matrix. Nat Rev Microbiol 2010;8:623–633.

39 O'Toole G, Kaplan HB, Kolter R: Biofilm formation as microbial development. Annu Rev Microbiol 2000;54:49–79.

40 Watnick P, Kolter R: Biofilm, city of microbes. J Bacteriol 2001;182:2675–2679.

41 Venketaraman V, Lin AK, Le A, Kachlany SC, Connell ND, Kaplan JB: Both leukotoxin and poly-*N*-acetylglucosamine surface polysaccharide protect *Aggregatibacter actinomycetemcomitans* cells from macrophage killing. Microb Pathog 2008;45:173–180.

42 Yamane K, Yamanaka T, Yamamoto N, Furukawa T, Fukushima H, Walker CB, Leung KP: A novel exopolysaccharide from a clinical isolate of *Prevotella nigrescens*: purification, chemical characterization and possible role in modifying human leukocyte phagocytosis. Oral Microbiol Immunol 2005;20:1–9.

43 Fukushima H, Moroi H, Inoue J, Onoe T, Ezaki T, Yabuuchi E, Leung K-P, Walker CB, Clark WB, Sagawa H: Phenotypic characteristics and DNA relatedness in *Prevotella intermedia* and similar organisms. Oral Microbiol Immunol 1992;7:60–64.

44 Yamanaka T, Furukawa T, Matsumoto-Mashimo C, Yamane K, Sugimori C, Nambu T, Mori N, Nishikawa H, Walker CB, Leung KP, Fukushima H: Gene expression profile and pathogenicity of biofilm-forming *Prevotella intermedia* strain 17. BMC Microbiol 2009;9:11.

45 Brunner J, Scheres N, El Idrissi NB, Deng DM, Laine ML, van Winkelhoff AJ, Crielaard W: The capsule of *Porphyromonas gingivalis* reduces the immune response of human gingival fibroblasts. BMC Microbiol 2010;10:5.

46 Davey ME, Duncan MJ: Enhanced biofilm formation and loss of capsule synthesis: deletion of a putative glycosyltransferase in *Porphyromonas gingivalis*. J Bacteriol 2006;188:5510–5523.

47 Yamaguchi M, Sato K, Yukitake H, Noiri Y, Ebisu S, Nakayama K: A *Porphyromonas gingivalis* mutant defective in a putative glycosyltransferase exhibits defective biosynthesis of the polysaccharide portions of lipopolysaccharide, decreased gingipain activities, strong autoaggregation, and increased biofilm formation. Infect Immun 2010;78:3801–3812.

48 Yeung MK: Molecular and genetic analyses of *Actinomyces* spp. Crit Rev Oral Biol Med 1999;10: 120–138.

49 Whittaker CJ, Klier CM, Kolenbrander PE: Mechanisms of adhesion by oral bacteria. Annu Rev Microbiol 1996;50:513–552.

50 Wu H, Fives-Taylor PM: Molecular strategies for fimbrial expression and assembly. Crit Rev Oral Biol Med 2001;12:101–115.

51 Chen P, Cisar JO, Hess S, Ho JT, Leung KP: Amended description of the genes for synthesis of *Actinomyces naeslundii* T14V type 1 fimbriae and associated adhesin. Infect Immun 2007;75:4181–4185.

52 Dorn BR, Leung K-P, Progulske-Fox A: Invasion of human oral epithelial cells by *Prevotella intermedia*. Infect Immun 1998;66:6054–6057.

53 Mishra A, Wu C, Yang J, Cisar JO, Das A, Ton-That H: The *Actinomyces oris* type 2 fimbrial shaft FimA mediates co-aggregation with oral streptococci, adherence to red blood cells and biofilm development. Mol Microbiol 2010;77:841–854.

54 Desvaux M, Khan A, Beatson SA, Scott-Tucker A, Henderson IR: Protein secretion systems in *Fusobacterium nucleatum*: genomic identification of Type 4 piliation and complete Type V pathways brings new insight into mechanisms of pathogenesis. Biochim Biophys Acta 2005;1713:92–112.

55 Tomich M, Planet PJ, Figurski DH: The *tad* locus: postcards from the widespread colonization island. Nat Rev Microbiol 2007;5:363–375.

56 Inoue T, Shingaki R, Sogawa N, Sogawa CA, Asaumi J, Kokeguchi S, Fukui K: Biofilm formation by a fimbriae-deficient mutant of *Actinobacillus actinomycetemcomitans*. Microbiol Immunol 2003;47:877–881.

57 Kaplan JB: Biofilm dispersal: Mechanisms, clinical implications, and potential therapeutic uses. J Dent Res 2010;89:205–218.
58 Dominiak DM, Nielsen JL, Nielsen PH: Extracellular DNA is abundant and important for microcolony strength in mixed microbial biofilms. Environ Microbiol 2011;13:710–721.
59 Stewart PS, Costerton JW: Antibiotic resistance of bacteria in biofilms. Lancet 2002;358:135–138.

Rudolf Gmür
Institute of Oral Biology, Section of Oral Microbiology and Immunology, University of Zürich
Plattenstrasse 11
CH–8032 Zürich (Switzerland)
Tel. +41 44 634 3306, E-Mail Rudolf.gmuer@zzm.uzh.ch

Kinane DF, Mombelli A (eds): Periodontal Disease. Front Oral Biol. Basel, Karger, 2012, vol 15, pp 17–40

Microbial Diversity and Interactions in Subgingival Biofilm Communities

Patricia I. Diaz

Division of Periodontology, Department of Oral Health and Diagnostic Sciences, School of Dental Medicine, The University of Connecticut Health Center, Farmington, Conn., USA

Abstract

The human subgingival environment is a complex environmental niche where microorganisms from the three domains of life meet to form diverse biofilm communities that exist in close proximity to the host. Bacteria constitute the most abundant, diverse and ultimately well-studied component of these communities with about 500 bacterial taxa reported to occur in this niche. Cultivation and molecular approaches are revealing the breadth and depth of subgingival biofilm diversity as part of an effort to understand the subgingival microbiome, the collection of microorganisms that inhabit the gingival crevices. Although these investigations are constructing a pretty detailed taxonomical census of subgingival microbial communities, including inter-subject and temporal variability in community structure, as well as differences according to periodontal health status, we are still at the front steps in terms of understanding community function. Clinical studies that evaluate community structure need to be coupled with biologically relevant models that allow evaluation of the ecological determinants of subgingival biofilm maturation. Functional characteristics of subgingival biofilm communities that still need to be clarified include main metabolic processes that support microbial communities, identification of keystone species, microbial interactions and signaling events that lead to community maturation and the relationship of different communities with the host. This manuscript presents a summary of our current understanding of subgingival microbial diversity and an overview of experimental models used to dissect the functional characteristics of subgingival communities. Future coupling of 'omics'-based approaches with such models will facilitate a better understanding of subgingival ecology opening opportunities for community manipulation.

The study of subgingival biofilms has been at the forefront of oral microbiology since traditional microscopic and culturing approaches revealed that shifts in bacterial community composition, in the susceptible host, are associated with periodontal inflammatory conditions [1, 2]. It is thus clear that the structure of health-associated communities is different from those in disease. However, the determinants of such ecological shifts are not completely understood. The first obvious difference between a host-compatible and a pathogenic biofilm is probably the biofilm biomass. Moore and Moore [3] reported that the total number of cultivable bacteria recovered after a single pass of a scaler instrument into a 1- to 2-mm deep gingival crevice is in the range of 10^3–10^4 cells, while the number of bacterial cells recovered from a site with mild-to-moderate periodontitis (4- to 5-mm deep gingival sulcus) ranges from 10^7 to 10^8 cells. Socransky and Haffajee [4] calculated the total numbers of microbial cells present in the subgingival crevices of healthy subjects to be 33.6×10^8 and of periodontitis subjects 174.3×10^8, a 10-fold difference. Moreover, differences between health- and disease-associated communities are also qualitative in nature, with shifts in community structure being the main factor associated with disease progression [5, 6]. For example, disease-associated microbial communities have been shown to be more diverse than health-associated communities with specific species displaying variable prevalence and abundance according to the health status of the gingival crevices [7, 8]. Furthermore, differences in community structure between healthy and periodontally affected sites are not only obvious at the species levels, but also at the organismal level, with some specific genotypes of single species displaying different prevalence in subjects with periodontal health and disease [9, 10]. Thus, it is imperative that we understand in fine detail the structure and functional characteristics of subgingival communities, including the ecological determinants that affect community maturation and ecological shifts if we are to develop diagnostic tools and preventive approaches to periodontal diseases. The first part of this manuscript summarizes our current understanding of subgingival community structure based on traditional culture-based and current molecular methods.

Subgingival biofilm communities inhabit a complex environmental niche. Microorganisms take advantage of the hard non-shedding root surfaces for attachment, and are subject to the permanent influx of gingival crevicular fluid (GCF)-containing host-derived macromolecules that serve as nutrients for the community, but also immune cells and molecules deployed as a defense by the host. Thus, microbial communities survive in the subgingival environment by the collective degradation of salivary and GCF host components, the establishment of food chains and the subversion of the immune response. Microbial metabolic interactions and inter-species communication are thus important aspects of community survival that could also be the target of antimicrobial strategies based on an ecological understanding of subgingival biofilms. Deep sequencing technology will allow us to capture community composition and development in vivo, including not only its phylogeny but also functional characteristics of these communities as revealed by metagenomic, metatranscriptomic, metaproteomic and meta-metabolomic approaches. However, cultivation of microbial species and establishment of in vitro laboratory models that reproduce the subgingival environment are still essential as we try to unravel and potentially interfere with the complex inter-species relationships that govern subgingival biofilm maturation. The second section of this chapter will discuss our understanding of subgingival community function from in vitro experimental studies that have used distinct models. It is proposed that the application of massively parallel sequencing technologies and whole-community proteomic and biochemical analysis to in vitro experimental models is necessary in order to gain insight into how subgingival microbes interact with other community members and with the host.

Subgingival Biofilm Diversity Revisited

Cultivation and the Advent of Culture-Independent Approaches

During the second half of the twentieth century, several research groups conducted methodical investigations into the diversity of subgingival biofilms revealing that these communities are composed of hundreds of species [3, 11]. Most of these investigations were carried out by cultivating subgingival biofilm samples from healthy and periodontally involved sites, followed by the identification of isolates according to their phenotypic characteristics. In one of the most comprehensive of these studies, Moore and Moore [3] sampled 300 subjects, obtaining 51,000 isolates that were classified into 509 different taxa. Of these taxa, 368 were detected more than once while the remaining were isolated only once. We could argue that the former represent the predominant culturable resident taxa of the subgingival crevice of humans while the latter are part of the rare bacterial flora, present only in very low numbers. Alternatively, the isolates seen only one time could have originated from difficult-to-culture species and their isolation numbers do not reflect their prevalence and abundance. Indeed, it has been evident for a long time that culturing methods are not sufficient to characterize the diversity of subgingival communities. One of the prime examples of this limitation is the observation by microscopic methods that spirochaetes are one of the most abundant morphotypes associated with gingivitis and periodontitis [12, 13]. However, since spirochetes are resistant to cultivation, most studies employing such methods have not revealed this taxon as a significant component of subgingival plaque. In order to select for spirochaetes, the Moore studies enriched cultures on selective media, followed by cultivation on membrane filters and semisolid medium [3]. Despite these efforts, no more than 7 different treponeme taxa could be recovered, while later molecular methods have revealed the presence of around 60 different species of treponemes in subgingival plaque of humans [12, 14, 15].

Our understanding of microbial ecology took a turn by the advent of 16S ribosomal RNA-based phylogenetic methods to estimate bacterial diversity and reconstruct prokaryotic relationships [16–18]. Since the late 1980s, microbial ecologists have been exploring the biodiversity of the microbial world by obtaining and sequencing rRNA from microorganisms present in their natural environment and comparing them to sequences from known organisms characterized in the laboratory. With the development of PCR, the amplification of small-subunit ribosomal RNAs from environmental samples followed by cloning and Sanger sequencing became common practice. The choice of rRNAs, key elements of the protein-synthesizing machinery, for analysis of natural microbial populations recognizes the fact that these ancient molecules are functionally homologous in all organisms [19]. Thus, some stretches in their nucleotide sequence are highly conserved, and can be conveniently used for alignment of sequences and for the design of universal primers to simultaneously amplify rRNA originating from multiple species in a given environmental sample [20]. Moreover, the variable regions in rRNA molecules can be used to discriminate between sequences – thus inferring the evolutionary relationships of the organisms. This was demonstrated by Carl Woese and his colleagues [16] who constructed the first comprehensive phylogeny of life on earth from studies of small-subunit rRNA partial sequences, revealing that life forms on earth consist of three primary lines of evolutionary descent: Eubacteria, Archaebacteria and Eukarya. It is remarkable that the subgingival environment harbors microorganisms from these three lines of descent. Bacteria dominate in terms of numbers and diversity in subgingival communities. However, the potential importance of Archaea has been highlighted by molecular surveys that revealed the presence of several *Methanobrevibacter* sp. in the subgingival plaque of approximately one third of patients with periodontal disease [21]. Similarly, eukaryotic microorganisms such as fungi are common inhabitants of several oral niches including the subgingival environment where their role in the overall community structure remains unclear [22, 23].

One of the most comprehensive characterizations of subgingival bacterial diversity based on culture-independent molecular methods was conducted by Paster et al. [15]. In this study, 31 subjects with different periodontal conditions were sampled. After amplification of the full-length 16S rRNA genes in the samples (approx. 1,550 bp), a total of 2,522 clones were obtained and sequenced, among which 347 species were identified. It was estimated that by increasing the sampling effort to 10,000 more clones, an additional 67 species would have been encountered, thus, estimating the total diversity in subgingival plaque to approx. 415 species. The phyla that contributed the highest number of species (the most diverse) were in descending order: *Firmicutes, Spirochaetes, Bacteroidetes, Proteobacteria, Actinobacteria* and *Fusobacteria.* Remarkably, about 40% of the clones constituted novel sequences, some from, to this day, uncultured phyla such as TM7 and SR1. Continued efforts by this group have clarified the taxonomic relationships originally published, as part of a broader effort to characterize the human oral microbiome, an international initiative sponsored in the United States by the National Institutes of Health that aims to take advantage of recently developed sequencing technologies to characterize the microbial communities that reside in the human body [24, 25].

The oral cavity constitutes one of the best-studied microbial niches. However, all techniques used throughout the last decades to characterize subgingival microbial diversity have advantages and shortcomings leading to potentially different results. Cultivation of microorganisms results in an under-representation of the microflora present at the site due to the resistance of about 40% of subgingival species to grow in the laboratory [15]. Targeted approaches such as DNA-DNA checkerboard hybridization provide high throughput and are semiquantitative allowing the evaluation of not only community structure measurements, such as species prevalence and relative abundance (proportion relative to the total microbial counts), but also absolute microbial load [26]. However, such techniques are not open-ended, and thus limit the analysis to a set of species (40 in most of checkerboard assays) [26, 27]. The use of 16S rRNA amplification with universal primers followed by cloning and Sanger sequencing, microarray hybridization of amplicons or direct sequencing using massively parallel sequencing technologies also suffers from bias due to discrepancies in PCR amplification of oral species by different primer sets [28, 29]. Moreover, the relative abundance of species can be determined, but the absolute microbial load would require real-time PCR to be applied for further sample analysis. Such absolute quantification may be an important measure in the study of subgingival communities in relation to periodontal disease since several in vitro and in vivo studies have conclusively demonstrated that absolute levels of microorganisms are directly related to the host response [30–32]. Furthermore, estimation of community structure by all molecular techniques relies on equally efficient lysis of all bacterial cell types in order to represent their original proportions in the natural environment, a process that differs depending on the lysis protocol used, as we and others have shown [33, 34].

The biggest challenge for microbial ecologists, irrespective of the technique used to evaluate microbial diversity, is the controversy around what constitutes a species [35]. In eukaryotic organisms, a species has been defined as a group of phenotypically similar organisms that are able to reproduce sexually, and thus maintain a certain degree of DNA homogeneity due to inter-organismal DNA recombination. Since bacteria and archaea reproduce asexually and are subject to horizontal gene transfer events, the boundaries to define a species are not clear. Thus, a bacterial or archaeal species has been traditionally defined based on a polyphasic approach involving phenotypic characterization coupled with methods to determine genomic similarity [35]. For example, isolates with >70% homology by genome DNA-DNA hybridization are considered to belong to the same species. rRNA sequence comparisons have also been shown to be useful in establishing species identity, and a sequence similarity of 97% in the full length of the 16S rRNA gene is commonly used as the cutoff to define a species. However, examples abound where these arbitrarily established cutoffs do not correspond to the phenotypic differentiation of species [36, 37]. Therefore, investigators using molecular techniques to describe microbial diversity use the terms 'phylotype' or operational taxonomic unit, rather than species, to indicate a group of organisms that share a certain level of similarity in their 16S rRNA sequence. The optimal cutoff to define operational taxonomic units also varies if only a portion of the 16S rRNA gene is sequenced, as occurs in most studies that use direct sequencing, since it has been shown that phylogenetic distances calculated using different variable regions across the 16S rRNA gene differ from those calculated using full-length sequences [38]. Thus, because of the above discussed reasons, oral microbiologists should consider the standardization of sample preparation procedures (such as lysis protocols), primers used for amplification as well as data analysis pipelines a priority, in order to make deep sequencing studies conducted by different groups comparable.

Most Abundant Taxa in Subgingival Biofilms

If we are to understand the drivers of subgingival biofilm formation, our aim should be an understanding of subgingival biofilm community function. It therefore seems an appropriate first step to characterize the metabolic processes and interactions of the most abundant components of the subgingival microbiota since these species would carry the bulk load of the metabolic activity in the community. A summary of abundance of subgingival biofilm taxa obtained via different approaches is presented in this section. Table 1 starts by presenting a summary of the relative abundance of bacterial genera in the subgingival environment in subjects with periodontal health, gingivitis and periodontitis as reported in the cultivation studies of Moore and Moore [3]. The most abundant genera found across samples were calculated by adding the reported relative abundance of individual species belonging to each genera in each periodontal health status category. If we consider only genera that constitute 5% or more of the isolates in at least one of the reported categories, it becomes evident that only three genera: *Actinomyces, Streptococcus* and *Fusobacterium* account for 46% of the total isolates across categories. If we add the three next most abundant genera, *Veillonella, Capnocytophaga* and *Parvimonas* (formerly *Peptostreptococcus*), it is seen that these 6 genera account for 56% of the total cultivable species in subgingival biofilms. Other important genera that appeared as at least 3% of the total isolates in any given sample included *Gemella, Prevotella, Selenomonas, Campylobacter* and *Eubacterium.* In the second row of table 1, the most abundant subgingival genera inferred from the data reported by Socransky and Haffajee [4] using DNA-DNA checkerboard hybridization are presented. Within the most abundant genera, out of 40 species evaluated, we encounter once again *Actinomyces, Fusobacterium, Capnocytophaga* and *Streptococcus*, a striking similarity to the results reported by cultivation. *Prevotella* and *Campylobacter*, which were within 3–5% of the isolates in the cultivation studies, appeared in the checkerboard in greater abundance. *Veillonella, Tannerella, Porphyromonas* and *Eubacterium* were also present at 3–5% of the counts reported by checkerboard. The results from the molecular open-ended approach used by Paster et al. [15] expanded the number of species encountered and although some genera, such as *Atopobium* and *Enterobacter*, now appeared within the most abundant, it is still evident that *Fusobacterium, Streptococcus*, *Prevotella* and *Veillonella* are present in high numbers confirming the cultivation and checkerboard results. Species grouped in the *Eubacterium* cluster were also abundant and included clones from the genera *Eubacterium, Parvimonas* and *Peptostreptococcus* that were grouped together as their taxonomy appeared unresolved in the phylogenetic trees reported. Other taxa that constituted at least 3% of the clones included the phylum *Synergistetes*, a

Table 1. Most abundant bacterial genera present in subgingival biofilms as revealed by different methods to evaluate microbial diversity

Method used	Sample characteristics	Reference number	Results	Note
Cultivation	14,403 isolates from subjects with periodontal health, gingivitis or chronic periodontitis; 30 bacterial isolates per site from 2 sites (60 isolates per subject) were considered	3	*Actinomyces:* 13–29% *Streptococcus:* 6–22% *Fusobacterium:* 3–17% *Veillonella:* 1–9% *Capnocytophaga:* 1–11% *Parvimonas:* 2–5%	data represent range of percent of isolates belonging to each genera in subjects with periodontal health, gingivitis and chronic periodontitis; only genera constituting ≥5% of the isolates in at least 1 subject category were taken into account; data from *Spirochaetes* were not included as enrichment cultures were used
DNA-DNA checkerboard hybridization	20,247 samples from 184 periodontally healthy subjects and 592 adult subjects with chronic periodontitis	4	*Actinomyces:* 19% *Prevotella:* 14% *Fusobacterium:* 12% *Capnocytophaga:* 8% *Campylobacter:* 7% *Streptococcus:* 6%	data represent percent of bacterial counts relative to total counts reported for 40 species tested; data from individual species were grouped by genera; the most abundant genera (≥5% of total counts) are reported
16S rRNA amplification, cloning and Sanger sequencing	1,917 clones were included in this analysis; samples were taken from the 4 deepest pockets in 31 subjects with periodontal health, acute necrotizing gingivitis and various forms of periodontitis	15	*Fusobacterium:* 17% *Streptococcus:* 11% *Atopobium:* 7% *Enterobacter:* 6% *Eubacterium* cluster: 6% *Prevotella:* 5% *Veillonella:* 5%	data represent the percentage of clones from each genus out of the total number of clones reported in phylogenetic trees; data from *Spirochaetes* were not included since specific primers were used; the most abundant genera (≥5% of total clones) are reported
	4,500 clones included in the analysis; 15 periodontally healthy subjects and 15 subjects with chronic periodontitis; samples taken from all mesial sites in healthy subjects and from 4 healthy and 4 periodontally involved sites in periodontitis subjects	39	*Selenomonas:* 15% *Streptococcus:* 14% *Veillonella:* 14% *Campylobacter:* 12% *Peptostreptococcus/ Parvimonas:* 11% *Eubacterium:* 6% *Dialister:* 5%	data represent percentage of clones from each genus out of the total number of clones reported; the most abundant genera (≥5% of total counts) are reported

cluster containing several *Neisseria*-like species and the genera *Selenomonas, Rothia, Dialister* and *Actinomyces. Porphyromonas, Campylobacter* and *Capnocytophaga* were each about 2% of the total clones. A similar methodological approach used by Kumar et al. [39], yielded somewhat different results with *Selenomonas* appearing as the most abundant genus followed by *Streptococcus, Veillonella, Campylobacter, Peptostreptococcus, Eubacterium* and *Dialister*. Other abundant genera reported by cultivation [3], checkerboard [4] and the molecular survey performed by Paster et al. [15], such as *Capnocytophaga, Fusobacterium, Prevotella, Atopobium* and *Actinomyces,* were less than 1% of the total clones in the study of Kumar et al. [39]. The difference between the Paster et al. [15] and Kumar et al. [39] molecular studies appears not to be due to primer bias as both studies used similar primers for initial amplification, but rather to sampling technique since samples were taken using curettes in the former study, while Kumar et al. [39] sampled the subgingival sites via paper points. These two sampling approaches may yield different results as the organisms most firmly attached to the tooth surface are probably underrepresented in paper point samples.

The summary from the studies presented makes it evident that understanding of microbial interactions and subgingival biofilm community function would benefit by first dissecting the inter-species interactions and metabolic networks of the most abundant genera. If we take a closer look into the studies presented, more specifically at the contribution of species from each genus to the overall community abundance, it is seen that in most cases, one or two species from each genus predominate and thus cultivated representative strains of such predominant species could be employed for in vitro investigations. Although great advances in this respect exist with the creation of poymicrobial biofilm models that incorporate some of the most commonly studied taxa – such as streptococci, *Actinomyces*, fusobacteria and *Veillonella* [40–42] that are also well studied in terms of their metabolic requirements [43–46] and interactions with the host [47] – certain genera such as *Capnocytophaga*, *Selenomonas*, *Eubacterium* and *Campylobacter* remain understudied from every aspect.

Least Abundant Taxa: Do They Play a Role?

An understanding of the major metabolic cycles that affect the energy fluxes, interspecies interactions and relationship with the host of the most abundant taxa is essential to unravel community function. However, it appears that among the least abundant taxa are in general those organisms traditionally associated to a 'pathogenic' biofilm. The red complex bacteria, *Tanerella forsythia, Treponema denticola* and *Porphyromonas gingivalis*, seem to hold the greatest association with periodontally involved sites [8]. Although these species demonstrate virulence factors tested in vitro and in vivo and are capable of deregulating the immune response, the forces driving the ecological shifts that promote their increased abundance in disease are not completely clear [48]. Most likely these microorganisms depend on interactions with the most abundant members of the community for survival and proliferation, a feature that can be exploited for the development of ecologically driven anti-biofilm approaches [49–51]. It has been hypothesized that a change in the environmental conditions, such as an increase in the amount and quality of GCF exudate due to inflammation promotes emergence of 'rare' species as major determinants of community structure [52]. Such environmental perturbation is expected to affect also major metabolic pathways in the community as these formerly 'rare' species, now abundant, may impact the energy flux of the ecosystem. For example, if we compare the subgingival communities of healthy sites to those of sites with periodontal disease, we may find that an increase in the proportion of species with potent endopeptidase activity such as *P. gingivalis* will increase the availability of peptides derived from host macromolecules [53, 54]. Such an increase in available nitrogen sources may shift the overall metabolic and virulence properties of the community as increased amino acid metabolism will also have an effect increasing the environmental pH due to the generation of organic acids and ammonia [55, 56]. *P. gingivalis* responds to more basic pH by upregulating gingipain activity [56], which in turn degrades more host-derived proteins and glycoproteins. Other community members such as fusobacteria may utilize such peptides for growth [43, 57] and also respond to fluxes in local pH [58].

Although this manuscript has focused on describing 'broad' characteristics of microbial communities, there is evidence that fine-scale heterogeneity in the structure of natural communities is also ecologically relevant [59]. The oral cavity is a prime example where closely related strains coexist. For example, *P. gingivalis* has been described as a species with a clonal population, where lateral gene transfer and recombination events have produced a myriad of strains with variable genotypic characteristics [60, 61]. Some of the differentiating regions have been mapped to operons for capsule polysaccharide [62], fimbria-encoding genes [9] and proteolytic enzyme genes [63]. Remarkably, one individual can harbor several *P. gingivalis* strains genotypically dissimilar [9, 64] that display dramatic differences in their nutritional requirements [65] and virulence potential [66, 67]. Such genetic variability has been linked to the existence of multi-copy mobile elements in *P. gingivalis*, such as insertion sequences, MITEs (miniature inverted-repeat transposable elements) and conjugative transposons [68]. The availability of closely related strains in a given environmental niche has been interpreted as a process that maintains diversity in the face of selection events, and thus confers resilience to the population. Co-existence of strains with slightly different metabolic and virulence potentials may ensure survival under selective pressures. In the case of *P. gingivalis*, such community perturbation events encompass daily oral hygiene practices, competition with other community members, and changes in immune status and salivary flow.

Other not so abundant microorganisms that occupy terminal positions in the food chain may also have an important role in subgingival communities by allowing an increase in the overall bacterial biomass. An example of such organisms are Archaebacteria which despite their relatively low proportion in the biofilm biomass, contribute to the overall metabolic structure by removing hydrogen from the environment, thus creating more thermodynamically favorable conditions for anaerobic bacteria [69]. Indeed, it has been shown that the proportion of Archaea, relative to the total prokaryotic rDNA, positively correlates to the severity of periodontal diseases, reflecting the complex nature of mature subgingival biofilm communities [21]. Moreover, syntrophic relationships between hydrogen-forming anaerobic bacteria and hydrogen-utilizing methanogens such as *Methanobrevibacter* sp. have been proposed to have an important role in intestinal microbial communities, and it would not be surprising that a similar interaction takes place in the subgingival environment.

Understanding Subgingival Biofilm Community Structure and Function in the Post-Genomic Era

Post-Sanger sequencing technologies have revolutionized our ability to obtain sequences from an environmental sample, since recently developed platforms such as 454 (454 Life Sciences/Roche Applied Sciences), Solexa (Illumina Incorporated) and SOLiD (Applied Biosciences) circumvent the cloning step, allowing the collection of tens of thousands of sequence tags from a single amplicon sample in only one run of the instrument. Since such depth of sequencing is probably not necessary to obtain a representative image of the diversity of subgingival biofilms in one subject, it is then possible to reduce costs and increase the sample throughput by utilization of multiple identifier tag sequences incorporated within the universal primers used in the initial amplification step. By using this approach, Costello et al. [70] evaluated the bacterial communities across different body sites, including the oral cavity, in 5 subjects. The oral samples were obtained by either an oral rinse procedure or by scrapping the tongue of the subjects. The most abundant genera in the oral cavity were *Streptococcus* (18.7%; Firmicutes), *Pasteurella* (15.9%; γ-Proteobacteria), *Veillonella* (13.9%; Firmicutes), *Prevotella* (13.6%; Bacteroidetes) and *Neisseria* (9.9%; β-Proteobacteria). No study, however, has used the advantages of this technology to evaluate subgingival biofilm diversity, although several recently funded NIH projects demonstrate that investigations of this nature are currently underway. Thus, 454 pyrosequencing and related highly parallel sequencing approaches promise to be techniques of choice for the evaluation of oral diversity in large number of individuals representing a range of oral health status, shedding light into the ecological shifts associated with different oral conditions including periodontal diseases. Moreover, with the reduction in cost of next-generation sequencing techniques, it is possible to envision studies in the near future that will use a metagenomic approach with direct shotgun sequencing of in situ subgingival microbial community DNA. Such an approach has been successfully used to characterize simple environmental or in-vitro-modeled microbial communities, and is deemed more accurate than sequencing of rRNA gene amplicons for assessment of taxonomic composition of a community as it avoids the bias introduced by PCR. Furthermore, a direct metagenomic sequencing approach could also reveal functional characteristics of microbial communities, as was demonstrated in a landmark study that characterized a simple environmental biofilm from an abandoned mine in California [71]. In silico predictions of metabolic pathways of this microbiome were generated from metagenomic data and later validated experimentally establishing the main metabolic roles for the dominant species in the community [72]. A similar approach has also been used to investigate microbial communities of the human and mouse intestine revealing important associations between community phylogenetic and gene structure, the metabolic potential of the community and obesity [73, 74]. Pioneering efforts that have used metagenomic techniques to characterize oral biofilms have revealed that about 3% of genes in dental plaque are related to resistance to antibiotics and toxic compounds, suggesting that the oral cavity is an important reservoir for antimicrobial resistance [75]. Furthermore, metagenomic studies have also been coupled to metatranscriptomic, metaproteomic and meta-metabolomic approaches to gain insights into community function at different levels. For example, Mahowald et al. [76] used all these methods to characterize a simplified model, developed in germ-free mice, of the human intestinal microbiome. In this study, mice were colonized with a representative strain from the two most abundant phyla in the intestine; *Firmicutes* and *Bacteroidetes,* represented by the species *Eubacterium rectale* and *Bacteroides thetaiotaomicron*, respectively. Interactions between these two species and their interaction with the host were evaluated after whole-genome transcriptional profiling of each species in mice colonized with either organism alone or in combination, under three dietary conditions. Data inferred from transcriptional studies were then verified by proteomic and biochemical assays of carbohydrate metabolism. This study elegantly shows the strategies used by each species to adapt to the presence of the other one. For example, *B. thetaiotaomicron* increases expression of polysaccharide utilization loci and signals the host to produce mucosal glycans that itself, but not *E. rectale,* can utilize. *E. rectale* on the other hand, diversifies its energy sources by increasing expression of selected amino acid transporters and using metabolic end products generated by *B. thetaiotaomicron* as energy sources. A change in the diet did not affect *B. thetaiotaomicron* colonization while *E. rectale* was significantly reduced in mice fed high-sugar diets. Since an intimate relationship exists between obesity, the intestinal microbiota and energy balance, understanding these ecological relationships has important implications in gastroenterology. This is an important study that demonstrates the feasibility of characterizing microbial interactions using relevant biological models. The periodontal field, with an already advanced knowledge of the metabolic requirements and virulence potential of a vast proportion of subgingival species, can benefit from the application of 'omic' approaches to reconstruct the microbial successions and ecological shifts associated with periodontitis. Moreover, computational biology advances now allow the reconstruction of metabolic networks from genome data in a reverse ecology approach [77]. Gaining insights from sequence data on the potential nutritional requirements of microorganisms may contribute to the design of appropriate media for the isolation of yet to be cultured taxa, an important step if we are to characterize the virulence potential

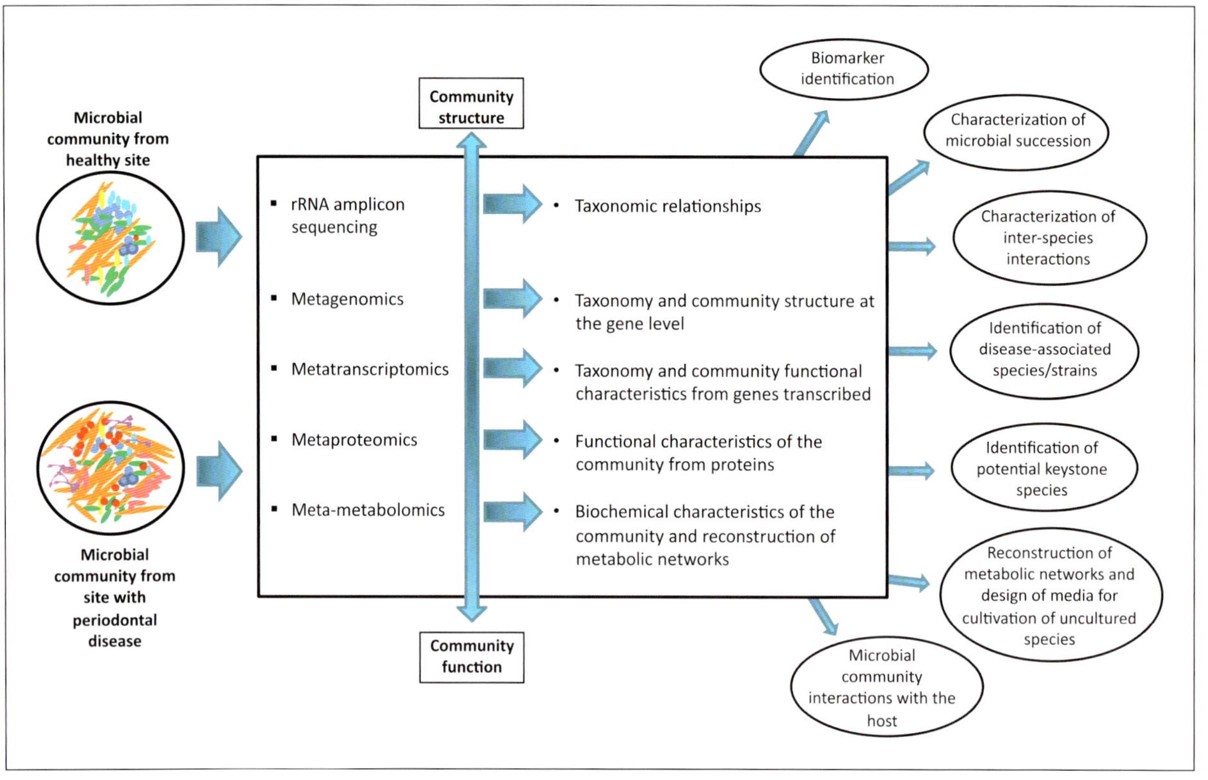

Fig. 1. Schematic representation of the use of 'omics' approaches to dissect the ecological determinants of health- and disease-associated subgingival biofilm communities. Different approaches and their outcomes are depicted in the middle panel. Methods closer to the top of the panel produce data related to community structure (species diversity and abundance), while methods closer to the bottom of the panel are more likely to reveal community characteristics related to their function. A combination of several of these techniques is necessary in order to understand microbial communities at all levels. Possible outcomes from such studies are depicted in the right circles.

of the uncultured portion of our subgingival flora. Metabolic networks can also inform us whether the nutritional requirements of one species are satisfied by the biosynthetic capacity of another, and thereby reveal important inter-species interactions. Figure 1 presents an overview of how 'omics'-based approaches can be used to inquire naturally occurring subgingival communities.

Dissecting the Ecological Determinants of Subgingival Biofilm Development in vitro

The understanding of microbial succession events that lead to biofilm maturation requires the development of in vitro experimental models that allow manipulation and testing of hypothesis regarding plaque formation in the laboratory. Ideal models need to be biologically relevant, reproducing the ecological environment of the oral cavity. Mouse models with humanized microbiota have proven useful in the study of the oral and intestinal microbiomes; however, such models carry high costs and are labor intensive [78, 79]. Thus, the creation of in vitro models that resemble the oral environment may be a more desirable end goal. Unanswered questions that could be answered in such models include: Which are the main metabolic processes that support the growth of subgingival communities? Which are the microbial interactions that facilitate the colonization and proliferation of late colonizers or species with more pathogenic potential? What is the role of coaggregation among bacterial species in biofilm maturation? Do any host components selectively enhance the proliferation of certain species in the biofilm? Do different strains from a particular species vary in their interactions with other members of the community? Can probiotic approaches control biofilm maturation? How does the immune response vary depending on the nature of the biofilm community challenge? In the next section, in vitro investigations that have tackled these questions will be discussed, including relevant models used.

Microbial Interactions among Subgingival Community Members

The process of biofilm formation is driven by repeatable microbial succession events in which some genera colonize nascent surfaces followed by the arrival of secondary and late colonizing species. The accretion of microbial biomass in the gingival crevices depends on the interaction among the different microbial species in the biofilm. These interactions occur at several levels, including physical contact, metabolic exchange, small-signal-molecule-mediated communication and exchange of genetic material [80, 81].

The specificity of physical interactions among oral biofilm bacteria was first described by Gibbons and Nygaard [82] and further dissected by the work of Paul Kolenbrander and colleagues [83–85] which evaluated the coaggregating ability of over a 1,000 oral bacteria isolates. Coaggregation is defined as the specific cell-to-cell recognition that occurs between genetically distinct cell types. During coaggregation, adhesins on the surface of one cell type recognize and bind to a complementary polysaccharide on the surface of another cell [86–88]. It has been proposed that coaggregation and close proximity facilitate bacterial mutualism, which influences biofilm development. Indeed, Jakubovics et al. [89] demonstrated that the expression of 23 genes was altered in *Streptococcus gordonii* following coaggregation with *Actinomyces oris*. Among the upregulated genes were seven genes in the *bfb* locus, which encodes phosphotransferase transport system components and associated β-glucosidases. The expression of these genes has also been reported to be essential for monospecies biofilm formation by *S. gordonii* [90]. Although, it is not clear which signal is responsible for this regulation, it appears that inter-species signaling mediated via coaggregation can trigger gene expression events that promote biofilm formation.

Close proximity has also been shown to be essential for inter-species interactions, even when the signal involves a diffusible molecule [91]. The work of Egland et al. [91] showed that a diffusible signal from *Veillonella atypica* mediates the upregulation of *S. gordonii amyB*, encoding α-amylase. Coaggregation was not shown to be essential for signaling, as the gene regulation event occurred in a closed environment where *V. atypica* and *S. gordonii* were separated by a dialysis membrane. However, coaggregation-mediated physical proximity among cells was demonstrated to be essential for upregulation of *S. gordonii amyB* in an open flow cell system where only *S. gordonii* cells juxtaposed to *V. atypica* upregulated *amyB*. This interaction has been speculated to benefit the growth of *Veillonella* as the upregulation of α-amylase may promote the mobilization of *S. gordonii*'s intracellular polysaccharide reserves, and thus increase its production of lactic acid, a primary fermentation substrate for *Veillonella*.

The interactions of early colonizing microorganisms with later colonizers are just beginning to unravel. Coaggregation may indeed have a role in driving microbial succession events during biofilm maturation. *Fusobacterium nucleatum*, an abundant component of oral biofilms has the ability to coaggregate with a diverse range of bacteria [84]. Thus, it has been postulated that *F. nucleatum* may promote the coadhesion of late colonizing species such as *Porphyromonas* sp. and *Treponema* sp. into the biofilm. Batch and continuous culture systems have been utilized to study these interactions. Continuous culture in a chemostat is a suitable in vitro model for the study of oral microorganisms, as it reproduces the submaximal growth rates at which microorganisms grow in natural conditions due to nutritional and environmental limitations [92]. The generation times of microorganisms in dental plaque are much lower than those from laboratory batch cultures, a situation that can be reproduced in the chemostat, which also allows close control of environmental factors such as pH, temperature, gas phase and E_h. The role of *F. nucleatum* as a species that promotes the growth of late colonizers has been tested in a two-stage continuous culture system where a polymicrobial culture was first established under anaerobic conditions and then passed to an aerated environment [93]. This study showed that removing *F. nucleatum* from the microbial consortia reduced the proportion of the gram-negative anaerobes *P. gingivalis* and *Prevotella nigrescens* in comparison to cultures containing *F. nucleatum*. Moreover, our group has also demonstrated that *F. nucleatum* contributes to generating a reducing and capnophilic environment that is necessary for the growth of *P. gingivalis* [94]. Interestingly, *P. gingivalis* and *F. nucleatum* did not coaggregate in suspension from anaerobic chemostat cultures, but were found in close physical contact under an oxygenated environment (fig. 2a–d). Thus, these studies support the role of physical proximity mediated by coaggregation as a means to establish environmentally favorable microenvironments for the growth of late colonizers.

The interaction of early colonizing streptococci with late colonizing species such as *P. gingivalis* has received much attention. It appears that *Streptococcus gordonii* can support the in vivo colonization [95] and in vitro biofilm growth of *P. gingivalis* [96].

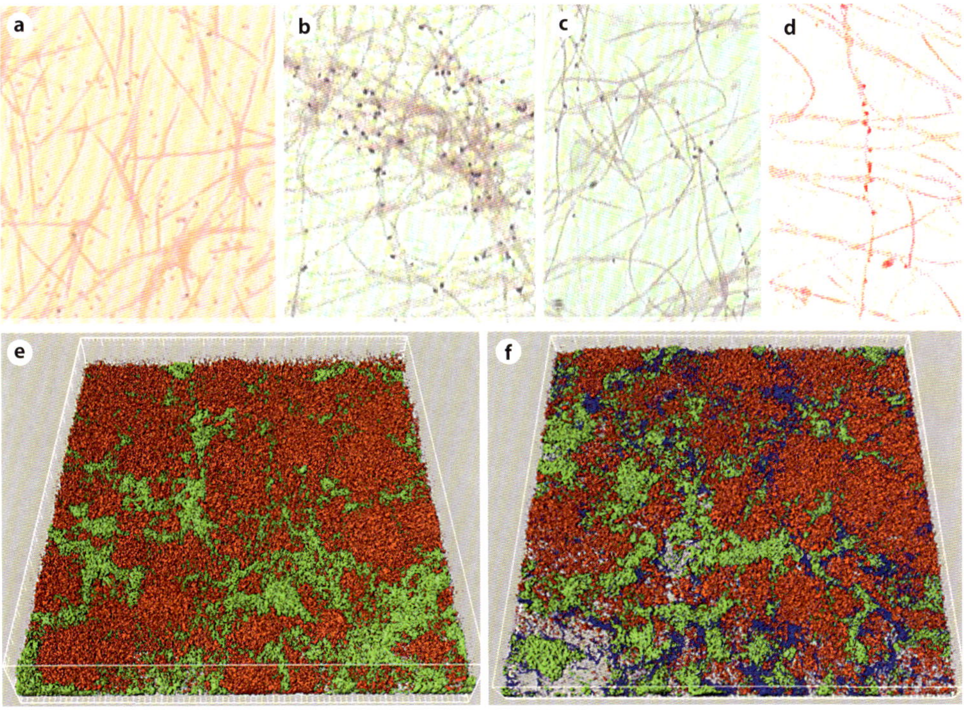

Fig. 2. Inter-species interactions in laboratory models of subgingival microbial communities. **a–d** Planktonic co-cultures of *Fusobacterium nucleatum* (rods) and *Porphyromonas gingivalis* (cocci) grown in a continuous culture chemostat system under different gaseous atmospheres, i.e. grown anaerobically (**a**) or in an aerated atmosphere (20% oxygen; **b–d**). Notice that in an anaerobic environment, most *P. gingivalis* cells do not coaggregate with *F. nucleatum*. However, in aerated cultures, a close physical proximity occurs. **e**, **f** 3D reconstructions of confocal laser scanning microscopy images of model subgingival biofilms grown in flow cells using saliva-supplemented medium depicting a biofilm of *Actinomyces oris* T14v (red) and *Veillonella* sp. PK1941 (green) grown for 16 h (**e**) or showing that *Fusobacterium nucleatum* ATCC 10953 (blue) can efficiently integrate itself into this community (**f**). *F. nucleatum* was inoculated after 4 h of biofilm growth by *Actinomyces* and *Veillonella* in order to mimic microbial succession, and the community was allowed to grow for an additional 12 h. Microorganisms in panels **e** and **f** were visualized after fluorescence in situ hybridization with genus-specific probes conjugated to Alexa fluorophores.

Interspecies signaling via autoinducer 2, a secondary metabolite product of the activated methyl cycle, demonstrated to act as a quorum sensing signal in many bacterial species, seems to be an important mediator of the *S. gordonii–P. gingivalis* interaction [97–99]. However, recent studies support the notion that the interaction of *P. gingivalis* with *S. gordonii* may not be favorable for *P. gingivalis* biofilm accretion as upon contact with *S. gordonii*, *P. gingivalis* upregulates a series of genes involved in regulatory networks that reduce adhesion and signaling, and thus decrease its ability to form a biofilm community with streptococci [98–100]. Indeed, the interaction of *P. gingivalis* with other streptococci has proven to be antagonistic. For example, *P. gingivalis* is able to form dual-species biofilms with *S. gordonii*, but not with streptococci such as *S. oralis*, *S. cristatus*, *S. intermedius* or *S. mutans* [41, 101]. Furthermore, it has become evident that *P. gingivalis* is able to sense extracellular arginine deiminase proteins produced by *S. cristatus* and *S. intermedius* and to respond by altering the expression of key surface structures required for colonization [102, 103]. Indeed, a negative correlation between the distribution of *S. cristatus* and *P. gingivalis* has been observed in subgingival plaque demonstrating that this antagonistic interaction has important implications for in vivo community maturation [104].

Perhaps one of the most important interactions occurring among oral bacteria is their ability to collectively degrade the main host macromolecules and utilize them as energy sources. The use of open flow cell systems where biofilms are allowed to develop on an abiotic surface using saliva as the only nutritional source has unequivocally demonstrated that oral bacteria consortia demonstrate better biofilm growth than single species [41, 42]. Model biofilm communities developed in a similar flow cell system are depicted in figure 2e, f. Chemostat studies have also shown that mucins can support the continuous growth of oral consortia and that the introduction of species with higher glycosidase and proteolytic activities in the consortia increases substrate utilization and biomass output [105]. Microbial nutrition in the subgingival environment, however, relies more on serum-derived proteins than on salivary components. A recent proteomic characterization of GCF revealed that its most abundant protein components are albumin, immunoglobulins and various keratins [106]. With an increase in GCF volume as a response to microbial accumulation in the gingival crevice, the greater availability of such macromolecules will favor the growth of non-saccharolytic highly proteolytic species. In this respect, *P. gingivalis* has been shown to be able to grow as a monoculture using bovine serum albumin as the sole carbon and nitrogen source [107], while streptococci and fusobacteria require synergistic interactions with other members of the biofilm community to grow on such substrates [108, 109]. Moreover, several *Treponema* sp. have been shown to require albumin for growth [109]. Thus, the study of in vitro microbial community growth in substrates similar to those encountered by microorganisms in vivo will advance our understanding of the overall metabolic interactions of communities.

Host Environment as a Determinant of Subgingival Biofilm Communities

In ecological theory, the niche is seen as a major determinant of community structure. Even within the oral cavity, differences among niches spatially contiguous to each other impact community structure in dramatic ways. For example, a comparison of the characteristics of the supra- and subgingival environments, two adjacent niches, reveals that differences in oxygen tension, redox potential, availability of fermentable carbohydrates and host-derived macromolecules, and pH at both sites are linked in a cause-and-effect way to the differences in the composition of the communities that inhabit them.

Furthermore, the subgingival niche changes as the community matures. The GCF, the main nutritional source for subgingival bacteria, is different from health to disease [106]. The growth of those bacteria associated with periodontal disease is enhanced by and increase in serum components, showing the close adaptation of subgingival communities to the host [110]. Moreover, inter-subject variability in community structure may not only be explained by differences in environmental factors, but also by the genetic background of the host. A recent study demonstrated that different subgingival cytokine profiles correlate with the existence of taxonomically different subgingival microbial communities [111]. Although it is always assumed that the microflora shapes the host response, it is tempting to reverse this thought and interpret the results from such studies as evidence that the host environment also determines the composition of subgingival communities. Thus, it is important to understand the contribution of inter-subject variability in the subgingival environment as a driver of subject-level differences in the phylogenetic structure of microbial communities. A study that has addressed this question in relation to the intestinal microbiome has demonstrated that individuality in the intestinal microflora is shaped by the host genotype [112].

Moreover, the epithelium has been regarded for many decades as a 'bystander' that would only serve as a barrier between the outside environment and the submucosa. This notion, however, has been challenged by the discovery of new epithelial functions that place the epithelium at the center stage of innate immunity [113]. The balance between a state of mucosal homeostasis and mucosal inflammation appears to be dependent on the contact of epithelial cells with different types of endogenous and exogenous ligands leading to activation of a series of cascades that ultimately result in a concerted tissue immune response [114]. A largely unexplored area, however, is how epithelial products play a role as immune effectors by regulating microbial growth in biofilms. Indeed, the host may modulate biofilm formation and maturation via secretion of epithelium- and neutrophil-derived peptides (e.g. defensins) [115, 116], cytokines [117] or other proteins such as lactoferrin [118, 119]. Secreted epithelial factors such as antimicrobial peptides could have a direct effect on the adjacent microflora. For example, human β-defensins are expressed in epithelial cells and exhibit broad-spectrum antimicrobial activity against oral microorganisms [115, 120]. The role of these peptides, however, in maintaining homeostasis in the oral cavity remains inconclusive, as the minimum inhibitory concentrations needed to kill most oral bacteria seem to be far greater than the concentration of human β-defensins detected in saliva [120, 121]. Thus, the significance of antimicrobial peptides in vivo has been proposed to be related to other mechanisms such as anti-inflammatory activity and inhibition of biofilm formation, rather than to a direct killing effect [115, 122]. Indeed, it has been demonstrated that the human antimicrobial peptide cathelicidin (LL-37) prevents in vitro biofilm formation by *Pseudomonas aeruginosa* at very low concentrations, far below those that kill or inhibit growth [116].

Other secreted host products could promote biofilm formation. In *Pseudomonas aeruginosa,* for example, virulence properties such as alginate production and surface hydrophobicity have been shown to be enhanced by macrophage secretory products containing a variety of cytokines [123]. *C. albicans* biofilm growth has also been shown to be enhanced by soluble products secreted by peripheral blood mononuclear cells [124]. Cytokines alone or in combination have also been reported to stimulate the in vitro planktonic growth of microorganisms such as *Escherichia coli* and *Mycobacterium avium* [117] and the biofilm growth of *Staphylococcus aureus* [125]. Thus, it is likely that the epithelium and its secreted products serve directly as an ecological determinant that modulates the composition and properties of adjacent biofilm communities. In this respect, the development of novel models that incorporate both a biofilm and a host component would greatly advance our ability to answer these questions [47].

Ecological Approaches to Biofilm Control

The study of ecological relationships among subgingival microorganisms will produce novel antimicrobial therapies. One possible approach involves the inhibition of adhesion interactions among microorganisms as a means to impede their colonization in the biofilm. The interaction between *S. gordonii* and *P. gingivalis* has been effectively inhibited by a synthetic peptide that targets the streptococcal adhesin SspB. The effectiveness of inhibiting this interaction has been demonstrated in a murine model of periodontitis where the peptide inhibited *P. gingivalis* colonization and bone loss in *P.-gingivalis/S.-gordonii*-infected mice [51]. On the other hand, signaling events may provide a broader target for anti-biofilm strategies. Since bacteria sense and respond to the presence of AI-2, not only produced by themselves but also by other species, and several studies have revealed that AI-2-mediated signaling is important for oral biofilm development [97, 126], the development of AI-2 antagonists seems a reasonable anti-biofilm strategy. In this respect, novel substances that act as AI-2 inhibitors are being evaluated against non-oral microorganisms such as *Vibrio* sp. and await testing in oral biofilm models [127]. Manipulation of microbial communities by use of probiotics has also been proposed [50]. Probiotics are known to inhibit the growth of 'undesirable' species while promoting the growth of beneficial microorganisms. Furthermore, certain probiotic bacteria may function not only by controlling microbial growth, but also by modulating the immune response [128]. Thus, ecological approaches to biofilm control will likely have a place in the future of therapeutic strategies for periodontal disease.

References

1 Socransky SS, Haffajee AD: Evidence of bacterial etiology: a historical perspective. Periodontol 2000 1994;5:7–25.

2 Loe H, Theilade E, Jensen SB: Experimental gingivitis in man. J Periodontol 1965;36:177–187.

3 Moore WEC, Moore LV: Periodontal microbiota in different clinical conditions. Periodontology 2000 1994;5:66–77.

4 Socransky SS, Haffajee AD: Periodontal microbial ecology. Periodontol 2000 2005;38:135–187.

5 Tanner AC, Haffer C, Bratthall GT, Visconti RA, Socransky SS: A study of the bacteria associated with advancing periodontitis in man. J Clin Periodontol 1979;6: 278–307.

6 Tanner A, Maiden MF, Macuch PJ, Murray LL, Kent RL Jr: Microbiota of health, gingivitis, and initial periodontitis. J Clin Periodontol 1998;25:85–98.

7 Colombo AP, Boches SK, Cotton SL, Goodson JM, Kent R, Haffajee AD, Socransky SS, Hasturk H, Van Dyke TE, Dewhirst F, Paster BJ: Comparisons of subgingival microbial profiles of refractory periodontitis, severe periodontitis, and periodontal health using the human oral microbe identification microarray. J Periodontol 2009;80:1421–1432.

8 Socransky SS, Haffajee AD, Cugini MA, Smith C, Kent RL Jr: Microbial complexes in subgingival plaque. J Clin Periodontol 1998;25:134–144.

9 Amano A, Nakagawa I, Okahashi N, Hamada N: Variations of *Porphyromonas gingivalis* fimbriae in relation to microbial pathogenesis. J Periodontal Res 2004;39:136–142.

10 Haubek D, Ennibi OK, Poulsen K, Poulsen S, Benzarti N, Kilian M: Early-onset periodontitis in Morocco is associated with the highly leukotoxic clone of *Actinobacillus actinomycetemcomitans*. J Dent Res 2001;80:1580–1583.

11 Dzink JL, Socransky SS, Haffajee AD: The predominant cultivable microbiota of active and inactive lesions of destructive periodontal diseases. J Clin Periodontol 1988;15:316–323.

12 Choi BK, Paster BJ, Dewhirst FE, Gobel UB: Diversity of cultivable and uncultivable oral spirochetes from a patient with severe destructive periodontitis. Infect Immun 1994;62:1889–1895.

13 Armitage GC, Dickinson WR, Jenderseck RS, Levine SM, Chambers DW: Relationship between the percentage of subgingival spirochetes and the severity of periodontal disease. J Periodontol 1982; 53:550–556.

14 Dewhirst FE, Tamer MA, Ericson RE, Lau CN, Levanos VA, Boches SK, Galvin JL, Paster BJ: The diversity of periodontal spirochetes by 16S rRNA analysis. Oral Microbiol Immunol 2000;15:196–202.

15 Paster BJ, Boches SK, Galvin JL, Ericson RE, Lau CN, Levanos VA, Sahasrabudhe A, Dewhirst FE: Bacterial diversity in human subgingival plaque. J Bacteriol 2001;183:3770–3783.

16 Fox GE, Stackebrandt E, Hespell RB, Gibson J, Maniloff J, Dyer TA, Wolfe RS, Balch WE, Tanner RS, Magrum LJ, Zablen LB, Blakemore R, Gupta R, Bonen L, Lewis BJ, Stahl DA, Luehrsen KR, Chen KN, Woese CR: The phylogeny of prokaryotes. Science 1980;209:457–463.

17 Lane DJ, Pace B, Olsen GJ, Stahl DA, Sogin ML, Pace NR: Rapid determination of 16S ribosomal RNA sequences for phylogenetic analyses. PNAS 1985;82:6955–6959.

18 Pace NR: A molecular view of microbial diversity and the biosphere. Science 1997;276:734–740.

19 Woese CR, Stackebrandt E, Macke TJ, Fox GE: A phylogenetic definition of the major eubacterial taxa. Syst Appl Microbiol 1985;6:143–151.

20 Amann RI, Binder BJ, Olson RJ, Chisholm SW, Devereux R, Stahl DA: Combination of 16S rRNA-targeted oligonucleotide probes with flow cytometry for analyzing mixed microbial populations. Appl Environ Microbiol 1990;56:1919–1925.

21 Lepp PW, Brinig MM, Ouverney CC, Palm K, Armitage GC, Relman DA: Methanogenic Archaea and human periodontal disease. PNAS 2004;101: 6176–6181.

22 Aas JA, Barbuto SM, Alpagot T, Olsen I, Dewhirst FE, Paster BJ: Subgingival plaque microbiota in HIV positive patients. J Clin Periodontol 2007;34:189–195.

23 Urzua B, Hermosilla G, Gamonal J, Morales-Bozo I, Canals M, Barahona S, Coccola C, Cifuentes V: Yeast diversity in the oral microbiota of subjects with periodontitis: *Candida albicans* and *Candida dubliniensis* colonize the periodontal pockets. Med Mycol 2008;46:783–793.

24 Dewhirst FE, Chen T, Izard J, Paster BJ, Tanner AC, Yu WH, Lakshmanan A, Wade WG: The human oral microbiome. J Bacteriol 2010;192:5002–5017.

25 Peterson J, Garges S, Giovanni M, McInnes P, Wang L, Schloss JA, Bonazzi V, McEwen JE, Wetterstrand KA, Deal C, Baker CC, Di Francesco V, Howcroft TK, Karp RW, Lunsford RD, Wellington CR, Belachew T, Wright M, Giblin C, David H, Mills M, Salomon R, Mullins C, Akolkar B, Begg L, Davis C, Grandison L, Humble M, Khalsa J, Little AR, Peavy H, Pontzer C, Portnoy M, Sayre MH, Starke-Reed P, Zakhari S, Read J, Watson B, Guyer M: The NIH Human Microbiome Project. Genome Res 2009;19: 2317–2323.

26 Socransky SS, Smith C, Martin L, Paster BJ, Dewhirst FE, Levin AE: 'Checkerboard' DNA-DNA hybridization. Biotechniques 1994;17:788–792.

27 Haffajee AD, Cugini MA, Tanner A, Pollack RP, Smith C, Kent RL Jr, Socransky SS: Subgingival microbiota in healthy, well-maintained elder and periodontitis subjects. J Clin Periodontol 1998;25: 346–353.

28 Kroes I, Lepp PW, Relman DA: Bacterial diversity within the human subgingival crevice. Proc Natl Acad Sci U S A 1999;96:14547–14552.

29 de Lillo A, Ashley FP, Palmer RM, Munson MA, Kyriacou L, Weightman AJ, Wade WG: Novel subgingival bacterial phylotypes detected using multiple universal polymerase chain reaction primer sets. Oral Microbiol Immunol 2006;21:61–68.

30 Vlachojannis C, Dye BA, Herrera-Abreu M, Pikdoken L, Lerche-Sehm J, Pretzl B, Celenti R, Papapanou PN: Determinants of serum IgG responses to periodontal bacteria in a nationally representative sample of US adults. J Clin Periodontol 2010;37:685–696.

31 Papapanou PN: Population studies of microbial ecology in periodontal health and disease. Ann Periodontol 2002;7:54–61.

32 Huang GT, Haake SK, Park NH: Gingival epithelial cells increase interleukin-8 secretion in response to *Actinobacillus actinomycetemcomitans* challenge. J Periodontol 1998;69:1105–1110.

33 Diaz PI, Chalmers NI, Rickard AH, Kong C, Milburn CL, Palmer RJ Jr, Kolenbrander PE: Molecular characterization of subject-specific oral microflora during initial colonization of enamel. Appl Environ Microbiol 2006;72:2837–2848.

34 Morgan JL, Darling AE, Eisen JA: Metagenomic sequencing of an in vitro-simulated microbial community. PLoS One 2010;5:e10209.

35 Gevers D, Cohan FM, Lawrence JG, Spratt BG, Coenye T, Feil EJ, Stackebrandt E, Van de Peer Y, Vandamme P, Thompson FL, Swings J: Opinion: Re-evaluating prokaryotic species. Nat Rev Microbiol 2005;3:733–739.

36 Kilian M, Poulsen K, Blomqvist T, Havarstein LS, Bek-Thomsen M, Tettelin H, Sorensen UB: Evolution of *Streptococcus pneumoniae* and its close commensal relatives. PLoS One 2008;3:e2683.

37 Christensen H, Nordentoft S, Olsen JE: Phylogenetic relationships of *Salmonella* based on rRNA sequences. Int J Syst Bacteriol 1998;48(part 2):605–610.

38 Schloss PD: The effects of alignment quality, distance calculation method, sequence filtering, and region on the analysis of 16S rRNA gene-based studies. PLoS Comput Biol 2010;6:e1000844.

39 Kumar PS, Griffen AL, Moeschberger ML, Leys EJ: Identification of candidate periodontal pathogens and beneficial species by quantitative 16S clonal analysis. J Clin Microbiol 2005;43:3944–3955.

40 Guggenheim M, Shapiro S, Gmur R, Guggenheim B: Spatial arrangements and associative behavior of species in an in vitro oral biofilm model. Appl Environ Microbiol 2001;67:1343–1350.

41 Periasamy S, Kolenbrander PE: Mutualistic biofilm communities develop with *Porphyromonas gingivalis* and initial, early, and late colonizers of enamel. J Bacteriol 2009;191:6804–6811.

42 Periasamy S, Kolenbrander PE: Central role of the early colonizer *Veillonella* sp. in establishing multispecies biofilm communities with initial, middle, and late colonizers of enamel. J Bacteriol 2010;192: 2965–2972.

43 Takahashi N, Sato T: Dipeptide utilization by the periodontal pathogens *Porphyromonas gingivalis, Prevotella intermedia, Prevotella nigrescens* and *Fusobacterium nucleatum*. Oral Microbiol Immunol 2002;17:50–54.

44 Takahashi N, Yamada T: Glucose and lactate metabolism by *Actinomyces naeslundii*. Crit Rev Oral Biol Med 1999;10:487–503.

45 Shah HN, Gharbia SE: Ecological events in subgingival dental plaque with reference to *Bacteroides* and *Fusobacterium* species. Infection 1989;17:264–268.

46 Rogers AH, Zilm PS, Gully NJ, Pfennig AL, Marsh PD: Aspects of the growth and metabolism of *Fusobacterium nucleatum* ATCC 10953 in continuous culture. Oral Microbiol Immunol 1991;6:250–255.

47 Guggenheim B, Gmur R, Galicia JC, Stathopoulou PG, Benakanakere MR, Meier A, Thurnheer T, Kinane DF: In vitro modeling of host-parasite interactions: the 'subgingival' biofilm challenge of primary human epithelial cells. BMC Microbiol 2009;9: 280.

48 Darveau RP: Periodontitis: a polymicrobial disruption of host homeostasis. Nat Rev Microbiol 2010;8: 481–490.

49 Kolenbrander PE, Palmer RJ Jr, Rickard AH, Jakubovics NS, Chalmers NI, Diaz PI: Bacterial interactions and successions during plaque development. Periodontol 2000 2006;42:47–79.

50 Marsh PD, Moter A, Devine DA: Dental plaque biofilms: communities, conflict and control. Periodontol 2000 2011;55:16–35.

51 Daep CA, Novak EA, Lamont RJ, Demuth DR: Structural dissection and in vivo effectiveness of a peptide inhibitor of *Porphyromonas gingivalis* adherence to *Streptococcus gordonii*. Infect Immun 2011;79:67–74.

52 Marsh PD: Are dental diseases examples of ecological catastrophes? Microbiology 2003;149(part 2):279–294.
53 Grenier D, Imbeault S, Plamondon P, Grenier G, Nakayama K, Mayrand D: Role of gingipains in growth of *Porphyromonas gingivalis* in the presence of human serum albumin. Infect Immun 2001;69: 5166–5172.
54 Grenier D, Mayrand D, McBride BC: Further studies on the degradation of immunoglobulins by black-pigmented *Bacteroides*. Oral Microbiol Immunol 1989;4:12–18.
55 Takahashi N: Acid-neutralizing activity during amino acid fermentation by *Porphyromonas gingivalis, Prevotella intermedia* and *Fusobacterium nucleatum*. Oral Microbiol Immunol 2003;18:109–113.
56 McDermid AS, McKee AS, Marsh PD: Effect of environmental pH on enzyme activity and growth of *Bacteroides gingivalis* W50. Infect Immun 1988; 56:1096–1100.
57 Grenier D: Effect of proteolytic enzymes on the lysis and growth of oral bacteria. Oral Microbiol Immunol 1994;9:224–228.
58 Zilm PS, Gully NJ, Rogers AH: Growth pH and transient increases in amino acid availability influence polyglucose synthesis by *Fusobacterium nucleatum* grown in continuous culture. FEMS Microbiol Lett 2002;215:203–208.
59 Hunt DE, David LA, Gevers D, Preheim SP, Alm EJ, Polz MF: Resource partitioning and sympatric differentiation among closely related bacterioplankton. Science 2008;320:1081–1085.
60 Chen T, Hosogi Y, Nishikawa K, Abbey K, Fleischmann RD, Walling J, Duncan MJ: Comparative whole-genome analysis of virulent and avirulent strains of *Porphyromonas gingivalis*. J Bacteriol 2004;186:5473–5479.
61 Frandsen EV, Poulsen K, Curtis MA, Kilian M: Evidence of recombination in *Porphyromonas gingivalis* and random distribution of putative virulence markers. Infect Immun 2001;69:4479–4485.
62 Laine ML, van Winkelhoff AJ: Virulence of six capsular serotypes of *Porphyromonas gingivalis* in a mouse model. Oral Microbiol Immunol 1998;13: 322–325.
63 Nadkarni MA, Nguyen KA, Chapple CC, DeCarlo AA, Jacques NA, Hunter N: Distribution of *Porphyromonas gingivalis* biotypes defined by alleles of the kgp (Lys-gingipain) gene. J Clin Microbiol 2004;42:3873–3876.
64 van Winkelhoff AJ, Rijnsburger MC, van der Velden U: Clonal stability of *Porphyromonas gingivalis* in untreated periodontitis. J Clin Periodontol 2008;35: 674–679.
65 Leduc A, Grenier D, Mayrand D: Comparative growth of *Porphyromonas gingivalis* strains in a defined basal medium. Anaerobe 1996;2:257–261.
66 Kuboniwa M, Inaba H, Amano A: Genotyping to distinguish microbial pathogenicity in periodontitis. Periodontol 2000 2010;54:136–159.
67 Neiders ME, Chen PB, Suido H, Reynolds HS, Zambon JJ, Shlossman M, Genco RJ: Heterogeneity of virulence among strains of *Bacteroides gingivalis*. J Periodontal Res 1989;24:192–198.
68 Naito M, Hirakawa H, Yamashita A, Ohara N, Shoji M, Yukitake H, Nakayama K, Toh H, Yoshimura F, Kuhara S, Hattori M, Hayashi T: Determination of the genome sequence of *Porphyromonas gingivalis* strain ATCC 33277 and genomic comparison with strain W83 revealed extensive genome rearrangements in *P. gingivalis*. DNA Res 2008;15:215–225.
69 Gibson GR, Cummings JH, Macfarlane GT: Competition for hydrogen between sulphate-reducing bacteria and methanogenic bacteria from the human large intestine. J Appl Bacteriol 1988;65: 241–247.
70 Costello EK, Lauber CL, Hamady M, Fierer N, Gordon JI, Knight R: Bacterial community variation in human body habitats across space and time. Science 2009;326:1694–1697.
71 Tyson GW, Chapman J, Hugenholtz P, Allen EE, Ram RJ, Richardson PM, Solovyev VV, Rubin EM, Rokhsar DS, Banfield JF: Community structure and metabolism through reconstruction of microbial genomes from the environment. Nature 2004;428: 37–43.
72 Tyson GW, Lo I, Baker BJ, Allen EE, Hugenholtz P, Banfield JF: Genome-directed isolation of the key nitrogen fixer *Leptospirillum ferrodiazotrophum* sp. *nov*. from an acidophilic microbial community. Appl Environ Microbiol 2005;71:6319–6324.
73 Gill SR, Pop M, Deboy RT, Eckburg PB, Turnbaugh PJ, Samuel BS, Gordon JI, Relman DA, Fraser-Liggett CM, Nelson KE: Metagenomic analysis of the human distal gut microbiome. Science 2006;312: 1355–1359.
74 Turnbaugh PJ, Ley RE, Mahowald MA, Magrini V, Mardis ER, Gordon JI: An obesity-associated gut microbiome with increased capacity for energy harvest. Nature 2006;444:1027–1031.
75 Xie G, Chain PS, Lo CC, Liu KL, Gans J, Merritt J, Qi F: Community and gene composition of a human dental plaque microbiota obtained by metagenomic sequencing. Mol Oral Microbiol 2010;25:391–405.

76 Mahowald MA, Rey FE, Seedorf H, Turnbaugh PJ, Fulton RS, Wollam A, Shah N, Wang C, Magrini V, Wilson RK, Cantarel BL, Coutinho PM, Henrissat B, Crock LW, Russell A, Verberkmoes NC, Hettich RL, Gordon JI: Characterizing a model human gut microbiota composed of members of its two dominant bacterial phyla. PNAS 2009;106:5859–5864.

77 Roling WF, Ferrer M, Golyshin PN: Systems approaches to microbial communities and their functioning. Curr Opin Biotechnol 2010;21:532–538.

78 Gibbons RJ, Socransky SS, Kapsimalis B: Establishment of human indigenous bacteria in germ-free mice. J Bacteriol 1964;88:1316–1323.

79 Faith JJ, Rey FE, O'Donnell D, Karlsson M, McNulty NP, Kallstrom G, Goodman AL, Gordon JI: Creating and characterizing communities of human gut microbes in gnotobiotic mice. ISME J 2010;4:1094–1098.

80 Kolenbrander PE, Palmer RJ Jr, Periasamy S, Jakubovics NS: Oral multispecies biofilm development and the key role of cell-cell distance. Nat Rev Microbiol 2010;8:471–480.

81 Kolenbrander PE, Egland PG, Diaz PI, Palmer RJ Jr: Genome-genome interactions: bacterial communities in initial dental plaque. Trends Microbiol 2005;13:11–15.

82 Gibbons RJ, Nygaard M: Interbacterial aggregation of plaque bacteria. Arch Oral Biol 1970;15:1397–1400.

83 Kolenbrander PE, London J: Adhere today, here tomorrow: oral bacterial adherence. J Bacteriol 1993;175:3247–3252.

84 Kolenbrander PE, Andersen RN, Moore LV: Coaggregation of *Fusobacterium nucleatum, Selenomonas flueggei, Selenomonas infelix, Selenomonas noxia,* and *Selenomonas sputigena* with strains from 11 genera of oral bacteria. Infect Immun 1989;57:3194–3203.

85 Kolenbrander PE: Surface recognition among oral bacteria: multigeneric coaggregations and their mediators. Crit Rev Microbiol 1989;17:137–159.

86 Cassels FJ, Hughes CV, Nauss JL: Adhesin receptors of human oral bacteria and modeling of putative adhesin-binding domains. J Ind Microbiol 1995;15:176–185.

87 Cisar JO, Sandberg AL, Abeygunawardana C, Reddy GP, Bush CA: Lectin recognition of host-like saccharide motifs in streptococcal cell wall polysaccharides. Glycobiology 1995;5:655–662.

88 Whittaker CJ, Klier CM, Kolenbrander PE: Mechanisms of adhesion by oral bacteria. Annu Rev Microbiol 1996;50:513–552.

89 Jakubovics NS, Gill SR, Iobst SE, Vickerman MM, Kolenbrander PE: Regulation of gene expression in a mixed-genus community: stabilized arginine biosynthesis in *Streptococcus gordonii* by coaggregation with *Actinomyces naeslundii*. J Bacteriol 2008;190:3646–3657.

90 Kilic AO, Tao L, Zhang Y, Lei Y, Khammanivong A, Herzberg MC: Involvement of *Streptococcus gordonii* beta-glucoside metabolism systems in adhesion, biofilm formation, and in vivo gene expression. J Bacteriol 2004;186:4246–4253.

91 Egland PG, Palmer RJ Jr, Kolenbrander PE: Interspecies communication in *Streptococcus gordonii-Veillonella atypica* biofilms: signaling in flow conditions requires juxtaposition. PNAS 2004;101:16917–16922.

92 Rogers AH, de Jong MH, Zilm PS, van der Hoeven JS: Estimation of growth parameters for some oral bacteria grown in continuous culture under glucose-limiting conditions. Infect Immun 1986;52:897–901.

93 Bradshaw DJ, Marsh PD, Watson GK, Allison C: Role of *Fusobacterium nucleatum* and coaggregation in anaerobe survival in planktonic and biofilm oral microbial communities during aeration. Infect Immun 1998;66:4729–4732.

94 Diaz PI, Zilm PS, Rogers AH: *Fusobacterium nucleatum* supports the growth of *Porphyromonas gingivalis* in oxygenated and carbon-dioxide-depleted environments. Microbiol 2002;148(part 2):467–472.

95 Slots J, Gibbons RJ: Attachment of *Bacteroides melaninogenicus* subsp. *asaccharolyticus* to oral surfaces and its possible role in colonization of the mouth and of periodontal pockets. Infect Immun 1978;19:254–264.

96 Lamont RJ, El-Sabaeny A, Park Y, Cook GS, Costerton JW, Demuth DR: Role of the *Streptococcus gordonii* SspB protein in the development of *Porphyromonas gingivalis* biofilms on streptococcal substrates. Microbiology 2002;148(part 6):1627–1636.

97 McNab R, Ford SK, El-Sabaeny A, Barbieri B, Cook GS, Lamont RJ: LuxS-based signaling in *Streptococcus gordonii*: autoinducer 2 controls carbohydrate metabolism and biofilm formation with *Porphyromonas gingivalis*. J Bacteriol 2003;185:274–284.

98 Simionato MR, Tucker CM, Kuboniwa M, Lamont G, Demuth DR, Tribble GD, Lamont RJ: *Porphyromonas gingivalis* genes involved in community development with *Streptococcus gordonii*. Infect Immun 2006;74: 6419–6428.

99 Maeda K, Tribble GD, Tucker CM, Anaya C, Shizukuishi S, Lewis JP, Demuth DR, Lamont RJ: A *Porphyromonas gingivalis* tyrosine phosphatase is a multifunctional regulator of virulence attributes. Mol Microbiol 2008;69:1153–1164.

100 Chawla A, Hirano T, Bainbridge BW, Demuth DR, Xie H, Lamont RJ: Community signalling between *Streptococcus gordonii* and *Porphyromonas gingivalis* is controlled by the transcriptional regulator CdhR. Mol Microbiol 2010;78:1510–1522.

101 Cook GS, Costerton JW, Lamont RJ: Biofilm formation by *Porphyromonas gingivalis* and *Streptococcus gordonii*. J Periodontal Res 1998;33:323–327.

102 Xie H, Lin X, Wang BY, Wu J, Lamont RJ: Identification of a signalling molecule involved in bacterial intergeneric communication. Microbiology 2007;153(part 10):3228–3234.

103 Christopher AB, Arndt A, Cugini C, Davey ME: A streptococcal effector protein that inhibits *Porphyromonas gingivalis* biofilm development. Microbiology 2010;156(part 11):3469–3477.

104 Wang BY, Wu J, Lamont RJ, Lin X, Xie H: Negative correlation of distributions of *Streptococcus cristatus* and *Porphyromonas gingivalis* in subgingival plaque. J Clin Microbiol 2009;47:3902–3906.

105 Bradshaw DJ, Homer KA, Marsh PD, Beighton D: Metabolic cooperation in oral microbial communities during growth on mucin. Microbiology 1994;140(part 12):3407–3412.

106 Bostanci N, Heywood W, Mills K, Parkar M, Nibali L, Donos N: Application of label-free absolute quantitative proteomics in human gingival crevicular fluid by LC/MS E (gingival exudatome). J Proteome Res 2010;9:2191–2199.

107 Milner P, Batten JE, Curtis MA: Development of a simple chemically defined medium for *Porphyromonas gingivalis*: requirement for alpha-ketoglutarate. FEMS Microbiol Lett 1996;140:125–130.

108 Homer KA, Beighton D: Synergistic degradation of bovine serum albumin by mutans streptococci and other dental plaque bacteria. FEMS Microbiol Lett 1992;69:259–262.

109 Van Horn KG, Smibert RM: Albumin requirement of *Treponema denticola* and *Treponema vincentii*. Can J Microbiol 1983;29:1141–1148.

110 ter Steeg PF, Van der Hoeven JS, de Jong MH, van Munster PJ, Jansen MJ: Enrichment of subgingival microflora on human serum leading to accumulation of *Bacteroides species, Peptostreptococci* and Fusobacteria. Antonie Van Leeuwenhoek 1987;53: 261–272.

111 Teles RP, Gursky LC, Faveri M, Rosa EA, Teles FR, Feres M, Socransky SS, Haffajee AD: Relationships between subgingival microbiota and GCF biomarkers in generalized aggressive periodontitis. J Clin Periodontol 2010;37:313–323.

112 Benson AK, Kelly SA, Legge R, Ma F, Low SJ, Kim J, Zhang M, Oh PL, Nehrenberg D, Hua K, Kachman SD, Moriyama EN, Walter J, Peterson DA, Pomp D: Individuality in gut microbiota composition is a complex polygenic trait shaped by multiple environmental and host genetic factors. PNAS 2010;107: 18933–18938.

113 Gribar SC, Richardson WM, Sodhi CP, Hackam DJ: No longer an innocent bystander: epithelial toll-like receptor signaling in the development of mucosal inflammation. Mol Med 2008;14:645–659.

114 Nochi T, Kiyono H: Innate immunity in the mucosal immune system. Curr Pharm Des 2006;12:4203–4213.

115 Dale BA, Fredericks LP: Antimicrobial peptides in the oral environment: expression and function in health and disease. Curr Issues Mol Biol 2005;7:119–133.

116 Overhage J, Campisano A, Bains M, Torfs EC, Rehm BH, Hancock RE: Human host defense peptide LL-37 prevents bacterial biofilm formation. Infect Immun 2008;76:4176–4182.

117 Wilson M, Seymour R, Henderson B: Bacterial perturbation of cytokine networks. Infect Immun 1998; 66:2401–2409.

118 Eberhard J, Drosos Z, Tiemann M, Jepsen S, Schroder JM: Immunolocalization of lactoferrin in healthy and inflamed gingival tissues. J Periodontol 2006;77:472–478.

119 Alugupalli KR, Kalfas S: Degradation of lactoferrin by periodontitis-associated bacteria. FEMS Microbiol Lett 1996;145:209–214.

120 Diamond G, Beckloff N, Ryan LK: Host defense peptides in the oral cavity and the lung: similarities and differences. J Dent Res 2008;87:915–927.

121 Ghosh SK, Gerken TA, Schneider KM, Feng Z, McCormick TS, Weinberg A: Quantification of human beta-defensin-2 and -3 in body fluids: application for studies of innate immunity. Clin Chem 2007;53:757–765.

122 Bowdish DM, Davidson DJ, Hancock RE: Immunomodulatory properties of defensins and cathelicidins. Curr Top Microbiol Immunol 2006; 306:27–66.

123 Mittal R, Sharma S, Chhibber S, Harjai K: Effect of macrophage secretory products on elaboration of virulence factors by planktonic and biofilm cells of *Pseudomonas aeruginosa*. Comp Immunol Microbiol Infect Dis 2006;29:12–26.

124 Chandra J, McCormick TS, Imamura Y, Mukherjee PK, Ghannoum MA: Interaction of *Candida albicans* with adherent human peripheral blood mononuclear cells increases *C. albicans* biofilm formation and results in differential expression of pro- and anti-inflammatory cytokines. Infect Immun 2007; 75:2612–2620.

125 McLaughlin RA, Hoogewerf AJ: Interleukin-1beta-induced growth enhancement of *Staphylococcus aureus* occurs in biofilm but not planktonic cultures. Microb Pathog 2006;41:67–79.

126 Rickard AH, Palmer RJ Jr, Blehert DS, Campagna SR, Semmelhack MF, Egland PG, Bassler BL, Kolenbrander PE: Autoinducer 2: a concentration-dependent signal for mutualistic bacterial biofilm growth. Mol Microbiol 2006;60:1446–1456.

127 Brackman G, Defoirdt T, Miyamoto C, Bossier P, Van Calenbergh S, Nelis H, Coenye T: Cinnamaldehyde and cinnamaldehyde derivatives reduce virulence in *Vibrio* spp. by decreasing the DNA-binding activity of the quorum sensing response regulator LuxR. BMC Microbiol 2008;8: 149.

128 Cosseau C, Devine DA, Dullaghan E, Gardy JL, Chikatamarla A, Gellatly S, Yu LL, Pistolic J, Falsafi R, Tagg J, Hancock RE: The commensal *Streptococcus salivarius* K12 downregulates the innate immune responses of human epithelial cells and promotes host-microbe homeostasis. Infect Immun 2008;76: 4163–4175.

Patricia I. Diaz, DDS, MS, PhD
Division of Periodontology, Department of Oral Health and Diagnostic Sciences
School of Dental Medicine, The University of Connecticut Health Center
263 Farmington Ave. MC 1710
Farmington, CT 06030–0001 (USA)
Tel. +1 860 679 3702, E-Mail pdiaz@uchc.edu

Kinane DF, Mombelli A (eds): Periodontal Disease. Front Oral Biol. Basel, Karger, 2012, vol 15, pp 41–55

Innate Cellular Responses to the Periodontal Biofilm

Manju Benakanakere · Denis F. Kinane

Department of Periodontics, School of Dental Medicine, University of Pennsylvania, Philadelphia, Pa., USA

Abstract

This chapter addresses the host responses to the microbial biofilm that constitutes the subgingival dental plaque. The host response to infection draws upon the innate, inflammatory and adaptive immune systems, whose role is to provide the appropriate response to the offending microorganisms. In some cases, this will be little or no response when encountering 'commensals', and in other cases a gradated response depending very much on the host's own determination of the pathogenic nature of the microbial insult: and herein lies the root of variation in host responses that govern individual susceptibility. In some individuals and with some bacteria this will be an innate-only response, others will need to invoke the inflammatory response, and yet others will require the adaptive immune response – be it cellular, humoral or both – to reduce or remove the challenge from the microbes. Of course these responses would be somewhat easier to predict with a single pathogen challenge, and become infinitely more complex as the biofilm increases in complexity. Oral infections, in particular gingival inflammation, originate from not just one but many microorganisms. This polymicrobial infection may result in chronic inflammation, which may lead to tissue destruction, as evident in chronic periodontitis. Although many organisms are present in the subgingival biofilm, interestingly, the putative pathogens associated with gingivitis and periodontitis may comprise very small fractions of the total biomass. An understanding of the interaction of structural and defensive host cells with the biofilm is pivotal to understanding periodontal disease etiology and to developing tailored therapeutics. Thus, this chapter addresses the main structural cells, i.e. epithelial cells, exposed to the biofilm.

Our body's immune system comprises multiple cell types whose primary function is to fight invading pathogens – such as bacteria, fungi, viruses, parasitic worms and other microorganisms – so as to protect and maintain immune homeostasis. The innate immune system provides the most immediate (and often a completely sufficient) response when cells encounter pathogens, and the adaptive immune system provides a delayed (but typically effective) sustained immunity. Oral infections, in particular gingival inflammation, originate from not just one but many microorganisms; this is termed 'polymicrobial infection' [1]. This type of infection results in chronic inflammation, which may lead to tissue destruction, as evident in chronic periodontitis. As described by Zijnge et al. [pp. 1–16] and Diaz [pp. 17–40], the microorganisms within the gingival crevice grow in aggregates as thin films called dental microbial biofilm or dental plaque. The dental biofilm is one of the most complex microbial communities found in the human body, and comprises more than 700 different bacterial species belonging to gram-positive and gram-negative organisms, protozoa, fungi and viruses [2, 3]. Although many organisms are present in the subgingival biofilm, interestingly, the putative pathogens associated with gingivitis and periodontitis may comprise very small fractions of the total biomass. Very little is known about the host immune response to the biofilm in general and, in particular, knowledge about dental biofilm and host immunity is very limited. This is mainly due to technical difficulties in performing definitive experiments, the varied structure and organization of biofilm communities, the fact that bioflims are somewhat of a recent concept (the last two decades), and that it is hard to pinpoint the actual organisms that initiate the disease process. It is well known that this plaque initiates inflammatory conditions, such as gingivitis, which if untreated lead to periodontitis [4]. Hence, the study of the composition and structure of this biofilm and its interaction with host cells is pivotal in understanding periodontal disease etiology to develop better therapeutics that can eliminate these microorganisms and prevent tissue damage.

A landmark article by Zijnge et al. [5] on the structure and architecture of biofilm on natural teeth using the fluorescent in situ hybridization technique shed light on the predominant bacterial microcolonies and their spatial arrangement within the plaque. Confocal laser scanning microscopy has confirmed that plaque has a open architecture, as suggested by the electron microscopy reports of channels traversing from the outside of the biofilm to the enamel surface [6–8]. Live/dead stains have suggested that bacterial viability may vary throughout the biofilm, with the most viable bacteria present in the central part of plaque and lining the voids and channels [6] – although this is still questionable, as reported elsewhere [9, 10]. This architecture, with a matrix comprising a wide range of exopolymers, creates a complex environment in terms of the penetration and distribution of molecules within plaque. There have been reports of uneven patterns of penetration of solutes in biofilm generated in situ in volunteers [11] and in an in-vitro-generated biofilm model [12]. The metabolism from biofilm organisms develops gradients in nutrients, pH, oxygen, etc. that are critical for bacterial growth. However, these gradients are not linear but exhibit heterogeneity, as reported for pH [13]. This environment encourages microorganisms to coexist in plaque biofilms, preventing it being a homogeneous environment. As revealed by the structure of the plaque biofilm on natural teeth [5], the presence of different species distributed in random microcolonies or pockets rather than forming layers supports the notion of one species of bacteria attaching to another.

Biofilm-Related Oral Diseases

Loe et al. [14] presented a positive association of biofilm to gingivitis as early as 1965. Further, Socransky and colleagues [15–19], Slots and Genco [20, 21], Tanner et al. [22] and recently Dewhirst et al. [23] identified several gram-negative bacterial species as putative pathogens in oral tissue destruction. Dewhirst et al. [23] adopted 16S rRNA sequencing that identified 619 taxa in 13 phyla, which can be found in the Human Oral Microbiome Database (www.homd.org). Another condition caused by oral microbial colonization is peri-implantitis. Mombelli et al. [24] reported the association of gram-negative bacteria by analyzing microbiota of failing osseointegrated implants and suggested *Bacteriodes* and *Fusobacterium* spp. may play important roles in initiating implant failure. Following this study, Lee et al. [25] also analyzed microbiota associated with healthy implants, and suggested that gram-negative bacteria were implicated in the failure of implants [26].

Innate Immunity to Oral Pathogens

The developments in the field of innate immunity over the past decade have been astounding, mainly due to the identification of pattern recognition receptors (PRRs) and related mechanistic cellular signaling pathways. Several of these PRRs have been identified and characterized as pathogen sensors, including Toll-like receptors (TLRs), Nod-like receptors, RIG-like receptors and dectins [27, 28]. More recently, cooperation between different receptors and their cellular signaling pathways in recognizing pathogens and mounting an innate immune response – including chemokine receptors, integrins, G-protein-coupled receptors and the complement system – has been elucidated, particularly to oral pathogens (fig. 1) [29–31]. In the context of polymicrobial infection, the immune response triggered may be due to several cellular signaling pathways being activated at the same time, and it is unlikely that a signaling pathway is activated in isolation. Hence, it becomes imperative to study these pathways together, instead of the reductionist approach to dissect and focus on a particular signaling pathway related to polymicrobial infection of the oral biofilm, as the biofilm represents a wide array of microorganisms.

TLRs are among the most studied PRRs, due to their role in detecting varied pathogen-associated molecular patterns [32, 33]. So far, 10 human TLRs have been identified with a specific set of ligands that they recognize [34, 35], except for TLR10 which is an orphan receptor and a specific ligand has yet to be discovered. TLR11 has been shown to be present in the human genome, but it does not translate into protein because of the presence of a stop codon in its open reading frame [36]. Specifically, TLR4 recognizes the lipopolysaccharide of gram-negative bacteria [37], and TLR2 forms heterodimers with TLR1 or TLR6 and recognizes peptidoglycan, lipopeptide and lipoproteins [38]. Double-stranded RNA (dsRNA) is recognized by TLR3 [39],

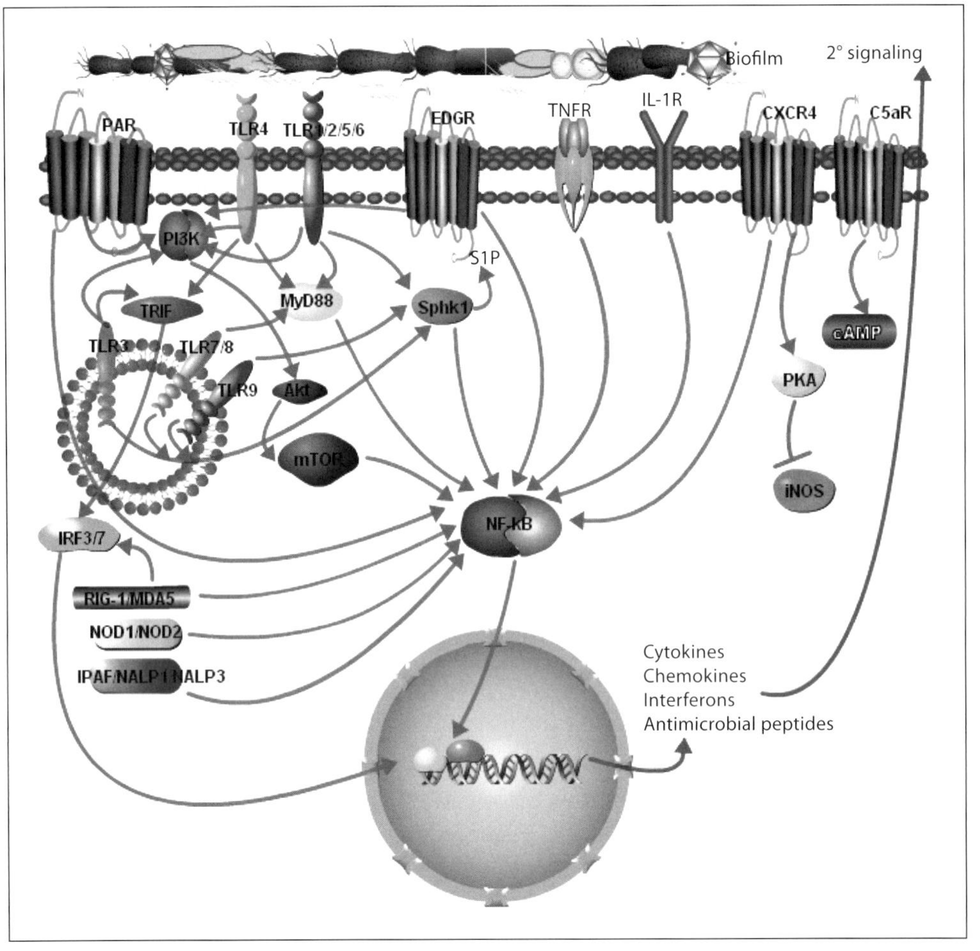

Fig. 1. Overview of a simplified hypothetical signaling cascade activated by TLRs and the TLR-crosstalk pathway by oral biofilm. Cellular signaling to oral biofilms may activate several unidentified pathways involving orphan G-protein-coupled receptors and intracellular PRRs and co-receptors leading to the immune response.

whereas TLR5 can detect bacterial flagellin [40]. TLR7 and TLR8 have been shown to recognize imidazoquinolines and single-stranded RNA [41, 42]. TLR9 recognizes bacterial and viral DNA with cytosine and guanine base pairing [43, 44]. PI3K is a heterodimeric protein that has been shown to be associated with TLR signaling [45–47]. These kinases can phosphorylate both lipids and proteins upon activation, resulting in the generation of PIP3 (phosphatidylinositol 3,4,5-triphosphate) from PIP2 (phosphatidylinositol 4,5-bisphosphate), recruiting signaling proteins like protein kinase B/Akt and subsequently activating NF-kB, leading to activation of inflammatory genes. Once these receptors, with the help of co-receptors or adaptor molecules, recognize pathogen-related biomolecules, several intracellular signaling events occur leading to the production of inflammatory cytokines, antimicrobial peptides, co-stimulatory molecules, type I interferons and chemokines to mount the innate immune response [48]. Recently, an association of this response with TLRs has been shown to induce antimicrobial peptide human β-defensin 2 in gingival epithelial cells [46].

The innate immune response is critical to facilitate the activation of antigen-presenting cells in shaping up sustained adaptive immune responses. Recently, the crosstalk between cytoplasmic PRRs, such as nucleotide-binding oligomerization domain (NOD)-like receptors (NLRs) and TLRs, has been identified [549–51]. NOD1 and 2, members of the NOD subfamily of NLR, recognize bacterial molecules involved in the metabolism of peptidoglycan, such as γ-D-Glu-mDAP (IE-DAP) and muramyl dipeptide, and induce the transcription of proinflammatory cytokines via the activation of NF-κB [52–54]. Flagellin can induce the Ipaf (ICE protease-activating factor)-dependent activation of caspase-1, whereas anthrax toxin and muramyl dipeptide can activate caspase-1 in a NALP1 (NACHT-LRR-PYD-containing protein 1)-dependent manner leading to the cleavage of pro-IL-1β to mature IL-1β [55]. Interestingly, NALP3 can recognize whole bacterium, RNA, uric acid or ATP to induce the caspase-1 activation required for IL-1β processing [56], and recently *Porphyromonas gingivalis* has also been shown to modulate NALP3 signaling [57, 58]. The NLRs include NALPs that can form inflammasomes (multi-protein complex within the cytoplasm), which are early events that activate proteolytic enzyme caspase-1. Caspase-1 has been shown to be critical in processing pro-IL-1β to its mature form [55, 56, 59]. Other than TLRs, cytosolic RIG-1 (retinoic-acid-inducible protein I, also known as Ddx58) and MDA5 (melanoma-differentiation-associated gene 5, also known as Ifih1 or Helicard) have also been shown to recognize viral dsRNA. The dsRNA recognition by RIG-1 and MDA5 leads to activation of IRF3 to induce interferon-β [60]. Very recently, the complement system has also been linked with TLR signaling [29, 61–63]. The complement cascade can be activated either by whole bacteria or by pathogen-related molecules that also activate TLRs and synergistically increase cAMP production. Downstream events in this crosstalk pathway include the activation of the cAMP-dependent protein kinase A inactivating glycogen synthase kinase-3β that impairs the iNOS-dependent killing of *P. gingivalis* in macrophages [29]. This crosstalk pathway has been shown to suppress TLR2-dependent nitric oxide in response to the pathogen utilizing a chemokine receptor, CXCR4 [64]. Recently, it has been shown that many of the pathogen-related molecules that activate TLRs can also activate sphingosine kinase 1 (Sphk-1) [46] and its activation leads to enhanced antimicrobial peptide human β-defensin 2 in gingival epithelial cells. Puneet et al. [65] identified Sphk-1 as a key regulator of proinflammatory cytokines in phagocytic cells upon polymicrobial infection. Polymicrobial infection seems to be much more complicated in terms of cellular activation than previously thought. Considering the diversity of microorganisms in biofilm colonization, the activation of multiple signaling pathways simultaneously seems to be relevant in polymicrobial-infection-related persistent inflammation, as in the case of periodontitis.

Use of planktonic bacteria or a single stimulus in various types of cells and animal models laid the foundation of our current understanding of cellular signaling related to the innate immune system. In nature however, the host encounters a wide array of microorganisms of diverse structure and organization. It is highly unlikely that an isolated cellular signaling pathway orchestrates the host's immune response to multiple organisms at a given time. It is more likely that multiple signaling pathways are activated simultaneously, depending on the nature and diversity of microorganisms and the ability of the host to fight back or succumb. Hence, it is very important to focus upon research using mixed cultures or biofilms to better understand the host-pathogen interaction in order to develop the next-generation therapeutics.

Bacterial biofilms are associated with number of diseases, including cystic fibrosis [66], endocarditis [67], periodontitis (most importantly as it can be readily identified) [15, 16, 18, 20], and various hospital-acquired diseases related to the use of catheters [68–70], heart valves [71] and orthopedic devices [72]. Hence, it is pivotal to study the innate immune response to pathogenic organisms in the form of a biofilm, and this allows us to movie one step closer to the natural situation. In this context, *Staphylococcus aureus* biofilms have been employed to study their effect on human leukocytes [73, 74]. However, leukocytes failed to kill or phagocytose *S. aureus* within the biofilm, contrary to planktonic bacteria, but secreted cytokines and type II interferon [74]. *Pseudomonas aeruginosa* biofilms have also been exploited to study the immune system [75]. When human neutrophils and macrophages were exposed to *P. aeruginosa* biofilm, the cells exhibited respiratory burst as well as phagocytosis and killing of the bacterial biofilm [75–77]. Closely following these observations, polymicrobial infection in periodontal disease and immune response was reported in a rat model [78]. However, this study was limited to mixing of plaktonically grown organisms *(P. gingivalis, Tannerella forsythia, Fusobacterium nucleatum* and *Treponema denticola)*. Nevertheless, the synergistic effect of the specific antibody IgG response has been detected with alveolar bone loss in polymicrobial infection [78].

Dental plaque has long been recognized as a complex polymicrobial biofilm [79]. Colonizing microorganisms in the gingival crevice initiate inflammation in the adjacent host gingival epithelium. Changes in microbiota from predominantly gram-positive facultative anaerobes to gram-negative anaerobic bacteria determine the biofilm maturity [80]. As the plaque accumulates, it induces inflammation in the adjacent host tissues, and over time the biofilm extends under the gum and down the root surface, creating a niche favoring the growth of anaerobes, such as *Spirochaetes* and *Bacteroides* [81]. The first and foremost aspect in unraveling the innate immune response to dental biofilm is to study epithelium-biofilm interactions, as the epithelium is the first line of defense and first encounters pathogens in the gingival crevice [4]. Until very recently, in vitro experiments to elucidate the mechanism of the innate immune system utilized human cell cultures stimulated with putative pathogenic bacteria that were grown in a planktonic state. Although evidence from such experiments has guided our initial understating of the innate immune response, the situation is far from what we can expect in vivo.

Biofilms are formed of either monospecies or multispecies biocenoses [82], and their eradication is more difficult than that of planktonic bacteria as they are highly resistant to antimicrobial agents and skilled in subverting the host's immune response [83]. Pathogenic biofilms are often associated with chronic inflammatory diseases (such as periodontitis) or chronic conditions that are difficult to treat (such as the colonization of urinary catheters and endotracheal tubes) [84, 85]. Biofilm or co-culture studies composed of 1–3 species have been reported in the past [86–88]; however, with hundreds of different bacteria present in the human oral cavity, an extensive biofilm study model will better elucidate the cellular responses triggered by bacteria that usually colonize with other species in vivo. Recently, there has been a new development with regard to cultivating multispecies biofilm and using it with primary gingival epithelial cells [79] – moving a few steps closer to the in vivo situation in understanding the host-pathogen interaction.

The bacterial species considered in this in vitro 'subgingival' biofilm model were carefully chosen based on their frequency in periodontal disease sites and relationship to 'red complex' organisms, as described by Sokransky et al. [89], and also that the species were easily cultivatable to permit various detection techniques, such as fluorescent in situ hybridization and immunofluorescence, in additional to species-specific single-cell detection assays. There was an attempt to develop a 9-species biofilm including widely accepted pathogenic microbiota [90]. In this study, 10 different oral microorganisms were used to generate in vitro biofilm *(Veillonella dispar, Campylobacter rectus, F. nucleatum, Prevotella intermedia, Actinomyces naeslundii, Streptococcus intermedius, Streptococcus oralis, T. forsythia, P. gingivalis* and *T. lecithinolyticum).* A detailed procedure on the development of multi-species biofilm can be found in the article by Guggenheim et al. [79].

Once the in vitro biofilm was developed, the human gingival epithelial cells (HGEC) cultures at the fourth passage were harvested and seeded at a density of 0.5×10^5 cells per well in a 6-well culture plate coated with type I collagen [46]. Biofilms on hydroxyapatite (HA) discs were grown as described [79] and frozen until further use. HA discs were defrosted and carefully revived by placing in fresh biofilm-medium (60% saliva, 10% human serum, 30% mFUM) or plain KSFM in an oxygen-containing atmosphere (90% N_2, 5% CO_2, 5% O_2) for 24 h. When the HGEC reached confluence (~90%), the HGEC were washed and then challenged for 4–24 h in antibiotic-free medium at 37°C in 5% CO_2 with 1 biofilm-carrying HA disc per well that had just been dipped 3 times in sterile normal saline solution. The co-culture set-up is illustrated in figure 2. Discs were placed on the ring support with the biofilm facing the HGEC monolayer. A 1-mm distance between the biofilm and the HGEC was maintained to allow fluid flow as per the literature on the gap between host tissue and biofilm attached to the tooth surface [91].

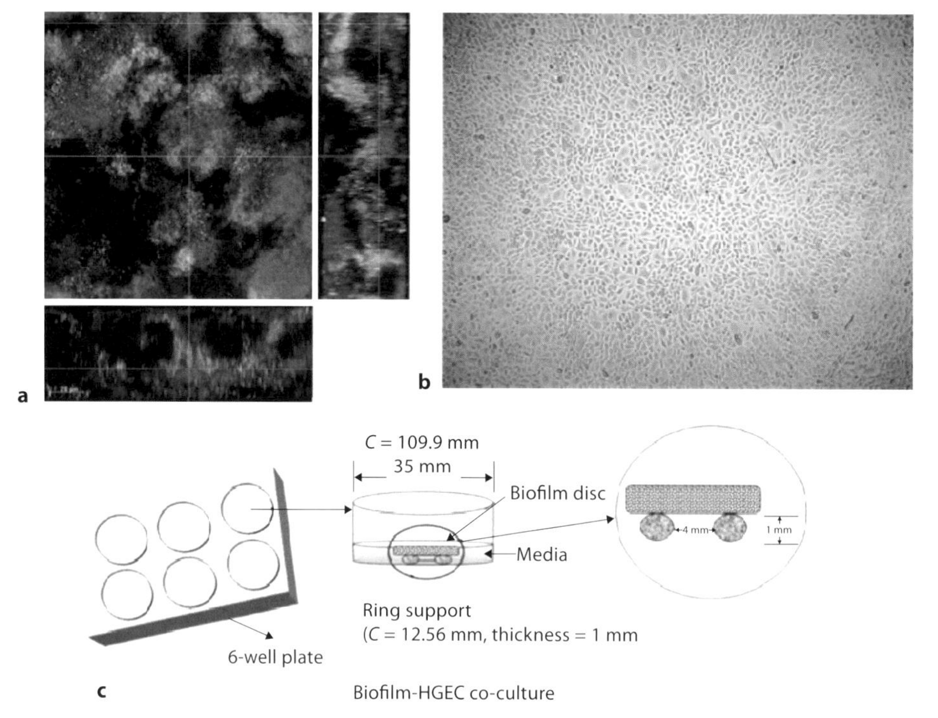

Fig. 2. In-vitro-developed biofilm on HA discs labeled with in situ fluorescent hybridization (**a**) and confluent HGECs used for co-culture experiment (**b**). Co-culture model showing the use of the 6-well plate and the ring support that fits into each well and between in vitro biofilm and HGECs (**c**). *C* = Circumference. Part of figure 2 was originally published by Biomed Central Ltd. [79].

Primary HGEC challenged with such 'subgingival' biofilms were able to induce a cytokine response [79]. Although the cytokine levels were higher compared to previously published reports [92, 93], the ability of biofilm to degrade the cytokine was lower but comparable to planktonic *P. gingivalis* challenge [93]. However, at 24 h, the levels of all cytokines diminished significantly, indicating the hydrolysis of the cytokines detected. This phenomenon could be similar to the observation from planktonic *P. gingivalis* challenge [93]. As in case of planktonic bacterial challenge, the biofilm-induced IL-6 and IL-8 degradation began immediately, and progressively increased to reach approximately 25% after 4 h of exposure, but only in the presence of intact biofilm (fig. 3). This suggests that the presence of biofilm bacteria is necessary for IL-6 and IL-8 degradation, while IL-1β was found to be more susceptible to degradation by a soluble component in the biofilm culture medium (fig. 3). Biofilm-induced cytokine degradation could be due to the gingipain activity of *P. gingivalis* or a protease of *T. forsythia* with similar trypsin-like activity. Previous studies have shown that *P. gingivalis* can induce cytokine degradation [93–95] that is mainly lysine-gingipain dependent.

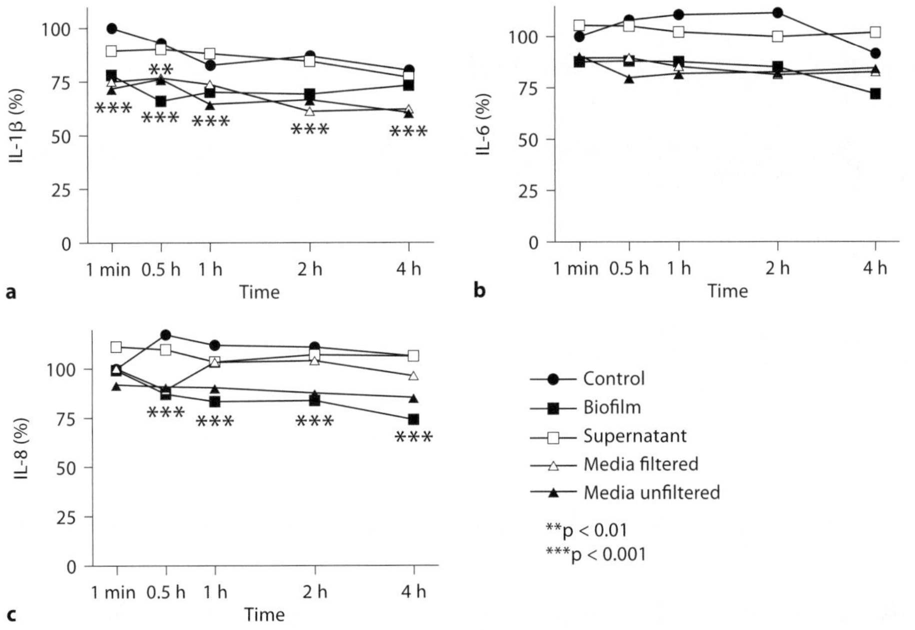

Fig. 3. Human gingival epithelial cells response to in vitro grown biofilm. Assays performed after 1 and 30 min and 1, 2 and 4 h of exposure to the biofilm showed that IL-1β (**a**), IL-6 (**b**) and IL-8 (**c**) degradation began immediately, and progressively increased to reach approximately 25% after 4 h.

The lower rate of the biofilm-induced cytokine degradation observed in the present study compared to that induced by *P. gingivalis* monocultures [93] is possibly dose-related or may be due to other organisms of the biofilm suppressing the expression of virulence factors, like gingipain by *P. gingivalis,* and this was not noted in planktonic bacterial challenge. Plaque-mediated cytokine degradation is a potential mechanism to explain the reduced levels of proinflammatory cytokines (IL-6) and chemokines (IL-8) found in the GCF of patients with periodontitis [96], as well as the GCF from sites with active periodontal disease or unresolved defects following treatment [97]. In addition to cytokine production, primary HGEC challenged in vitro with a 'subgingival' biofilm exhibited apoptosis [79]. This is a potential mechanism to explain the apoptosis that is observed in the gingiva at sites of chronic bacterially induced inflammation [98], particularly among the cells of the junctional epithelium and the fibroblasts and leucocytes of the connective tissue [98, 99]. The high proportions of HGEC undergoing apoptosis in the described co-culture system, as well as the short exposure time that is necessary for the initiation of the apoptotic process, suggest that the in vitro 'subgingival' biofilm is a highly pathogenic entity, and possibly more pathogenic than single bacterial strains. This model demonstrates that biofilms mimicking subgingival plaque can be established reproducibly in vitro and when exposed in co-culture to primary human epithelial cells, biofilms induce cellular apoptosis and stimulate the production of proinflammatory primary and secondary cytokines. In contrast to planktonic challenge, the responses of epithelial cells to biofilm were more robust than previously reported [92, 93]. This model also points out two key aspects of the interactions taking place at the dental plaque:periodontal tissue interphase, namely the triggering of a strong proinflammatory host cell response and the simultaneous microbial subversion of host defense mechanisms, including cellular apoptosis.

Role of Innate Immune Cells Triggering Adaptive Immune Responses to Polymicrobial Infection

The role of innate immune cells is generally to detect microbes and maintain host-microbial immune homeostasis and to induce antimicrobial defense mechanisms. Innate immune cells like epithelial cells, fibroblasts, dendritic cells, macrophages and neutrophils act as the first line of defense against invading pathogens and mount an adaptive immune response by antigen-presenting cells (dendritic cells and macrophages). The adaptive immune system is composed of T and B lymphocytes that become activated upon antigen presentation [100]. However, the specific mechanism linking innate to adaptive immune response is not fully understood. Although there have been studies that indicate the activation of adaptive immune responses in the absence of TLR function [35, 101], in general, polymicrobial sepsis has been show to activate several members of TLRs in antigen-presenting cells [102], which can engulf antigens to form phagosome-endosome that recruits and processes antigens to MHC class II to activate T cells [100]. This process has been shown to occur with the help of costimulatory molecules that are induced by TLRs. Endothelial cells have been shown to uptake *P. gingivalis* by autophagy [103] and this also has been shown to activate TLR2 [104, 105]. Autophagy has been linked to MHC class II antigen presentation [106, 107]. In this case, *P. gingivalis* may induce co-stimulatory molecules mediating antigen presentation, thereby connecting the innate immune response to adaptive immunity. TLRs are well-characterized receptors that can trigger activation of adaptive immune responses of several effector classes, including immunoglobulin M (IgM), IgG and IgA antibody responses and T helper cell 1 (TH1) and CD4+ and CD8+ T cell responses [108]. NLRs and RLRs can also activate adaptive immune responses [109–111]. TLRs can activate sphk-1 and activated sphk-1 phosphorylates S1P [46]. The S1P produced can regulate trafficking of lymphocytes and the migration of T cells and B cells. In addition, it also enhances production of IgE and IgE-mediated release of allergic mediators from mast cells [112]. In the context of oral biofilm, it becomes hard to distinguish the mechanism for triggering the adaptive immune response from the innate immune response because it is very difficult to discriminate between commensals and pathogens – mainly due to the lack of proteomic studies on biofilm. It can be discriminated by virulence factors from red complex bacteria, such as fimbriae, endotoxins and proteases, but this may be arbitrary and mostly conditional (i.e. based on the host immune response).

Future Perspectives

The study of biofilms and host interactions is pivotal in developing the next generation of therapeutics. In this regard, the present understanding of structure and organization of biofilm and its concomitant immune response is limited, despite the fact that dental plaque has been well known as a biofilm and that gingivitis and periodontitis have been recognized as polymicrobial infections for over two decades. Current research on the dental biofilm is limited to the identification of a number of microorganisms present in the microflora and its organization, structure, intermicrobial communication and formation. However, there is a need for high-throughput proteomic, lipidomic and metabolomic analyses of biofilm to understand the differential expression of virulence factors within the biofilm. There is also a need to develop in vitro biofilms for experiments that would comprise known bacteria, yeast, parasitic worms and viruses, so as to mimic in vivo biofilms, and to use these with in vitro or in vivo experiments on animal models to better understand the immune response. In addition, in vitro modeling using multiple human cell types (fibroblasts, macrophages, dendritic cells, neutrophils, etc.) may shed more light on the disease etiology. These aspects pose significant challenges to the research community and may require multicenter international research collaborations, but are truly the way forward in our understanding and management of dental diseases.

Acknowledgements

We thank Drs. Bernhard Guggenheim, Panagiota Stathopoulou and Johnah Galicia for their contribution to biofilm-host interaction study. This work was supported by United States Public Health Service, National Institutes of Health, NIDCR grant DE017384 to D.F.K.

References

1 Kinane DF, Hajishengallis G: Polymicrobial infections, biofilms, and beyond. J Clin Periodontol 2009; 36:404–405.

2 Lamont RJ, Jenkinson HF: Life below the gum line: pathogenic mechanisms of *Porphyromonas gingivalis*. Microbiol Mol Biol Rev 1998;62:1244–1263.

3 Paster BJ, Olsen I, Aas JA, Dewhirst FE: The breadth of bacterial diversity in the human periodontal pocket and other oral sites. Periodontol 2000 2006;42:80–87.

4 Kinane DF, Galicia JC, Gorr SU, Stathopoulou PG, Benakanakere M: *P. gingivalis* interactions with epithelial cells. Front Biosci 2008;13:966–984.

5 Zijnge V, van Leeuwen MB, Degener JE, et al: Oral biofilm architecture on natural teeth. PLoS One 2010;5:e9321.
6 Auschill TM, Arweiler NB, Netuschil L, Brecx M, Reich E, Sculean A: Spatial distribution of vital and dead microorganisms in dental biofilms. Arch Oral Biol 2001;46:471–476.
7 Wood SR, Kirkham J, Marsh PD, Shore RC, Nattress B, Robinson C: Architecture of intact natural human plaque biofilms studied by confocal laser scanning microscopy. J Dent Res 2000;79:21–27.
8 Zaura-Arite E, van Marle J, ten Cate JM: Conofocal microscopy study of undisturbed and chlorhexidine-treated dental biofilm. J Dent Res 2001;80:1436–1440.
9 Gelle MP, Jacquelin LF, Choisy C: Compared viability of planctonic bacteria and bacteria in biofilms by flow cytometry (in French). Ann Pharm Fr 2003;61:243–252.
10 Hope CK, Wilson M: Measuring the thickness of an outer layer of viable bacteria in an oral biofilm by viability mapping. J Microbiol Methods 2003;54:403–410.
11 Robinson C, Kirkham J, Percival R, et al: A method for the quantitative site-specific study of the biochemistry within dental plaque biofilms formed in vivo. Caries Res 1997;31:194–200.
12 Thurnheer T, Gmur R, Shapiro S, Guggenheim B: Mass transport of macromolecules within an in vitro model of supragingival plaque. Appl Environ Microbiol 2003;69:1702–1709.
13 Vroom JM, De Grauw KJ, Gerritsen HC, et al: Depth penetration and detection of pH gradients in biofilms by two-photon excitation microscopy. Appl Environ Microbiol 1999;65:3502–3511.
14 Loe H, Theilade E, Jensen SB: Experimental gingivitis in man. J Periodontol 1965;36:177–187.
15 Haffajee AD, Socransky SS, Ebersole JL, Smith DJ: Clinical, microbiological and immunological features associated with the treatment of active periodontosis lesions. J Clin Periodontol 1984;11:600–618.
16 Haffajee AD, Socransky SS, Goodson JM: Periodontal disease activity. J Periodontal Res 1982;17:521–522.
17 Socransky SS: Microbiology of plaque. Compend Contin Educ Dent 1984;(suppl 5):S53–S56.
18 Socransky SS, Haffajee AD, Goodson JM, Lindhe J: New concepts of destructive periodontal disease. J Clin Periodontol 1984;11:21–32.
19 Socransky SS, Tanner AC, Goodson JM, et al: An approach to the definition of periodontal disease syndromes by cluster analysis. J Clin Periodontol 1982;9:460–471.
20 Slots J, Genco RJ: Black-pigmented *Bacteroides* species, Capnocytophaga species, and Actinobacillus actinomycetemcomitans in human periodontal disease: virulence factors in colonization, survival, and tissue destruction. J Dent Res 1984;63:412–421.
21 Genco RJ, Slots J: Host responses in periodontal diseases. J Dent Res 1984;63:441–451.
22 Tanner AC, Paster BJ, Lu SC, et al: Subgingival and tongue microbiota during early periodontitis. J Dent Res 2006;85:318–323.
23 Dewhirst FE, Chen T, Izard J, et al: The human oral microbiome. J Bacteriol 2010;192:5002–5017.
24 Mombelli A, van Oosten MA, Schurch E Jr, Land NP: The microbiota associated with successful or failing osseointegrated titanium implants. Oral Microbiol Immunol 1987;2:145–151.
25 Lee KH, Maiden MF, Tanner AC, Weber HP: Microbiota of successful osseointegrated dental implants. J Periodontol 1999;70:131–138.
26 Sbordone L, Bortolaia C: Oral microbial biofilms and plaque-related diseases: microbial communities and their role in the shift from oral health to disease. Clin Oral Investig 2003;7:181–188.
27 Medzhitov R: Innate immunity: quo vadis? Nat Immunol 2010;11:551–553.
28 Medzhitov R: Inflammation 2010:new adventures of an old flame. Cell 2010;140:771–776.
29 Wang M, Krauss JL, Domon H, et al: Microbial hijacking of complement-toll-like receptor crosstalk. Sci Signal 2010;3:ra11.
30 Eskan MA, Rose BG, Benakanakere MR, Lee MJ, Kinane DF: Sphingosine 1-phosphate 1 and TLR4 mediate IFN-beta expression in human gingival epithelial cells. J Immunol 2008;180:1818–1825.
31 Eskan MA, Rose BG, Benakanakere MR, et al: TLR4 and S1P receptors cooperate to enhance inflammatory cytokine production in human gingival epithelial cells. Eur J Immunol 2008;38:1138–1147.
32 Medzhitov R, Preston-Hurlburt P, Kopp E, et al: MyD88 is an adaptor protein in the hToll/IL-1 receptor family signaling pathways. Mol Cell 1998;2:253–258.
33 Yang RB, Mark MR, Gray A, et al: Toll-like receptor-2 mediates lipopolysaccharide-induced cellular signalling. Nature 1998;395:284–288.
34 Kawai T, Akira S: TLR signaling. Semin Immunol 2007;19:24–32.
35 Koyama S, Ishii KJ, Kumar H, et al: Differential role of TLR- and RLR-signaling in the immune responses to influenza A virus infection and vaccination. J Immunol 2007;179:4711–4720.
36 Zhang D, Zhang G, Hayden MS, et al: A toll-like receptor that prevents infection by uropathogenic bacteria. Science 2004;303:1522–1526.

37 Poltorak A, He X, Smirnova I, et al: Defective LPS signaling in C3H/HeJ and C57BL/10ScCr mice: mutations in Tlr4 gene. Science 1998;282:2085–2088.

38 Takeuchi O, Sato S, Horiuchi T, et al: Cutting edge: role of Toll-like receptor 1 in mediating immune response to microbial lipoproteins. J Immunol 2002;169:10–14.

39 Alexopoulou L, Holt AC, Medzhitov R, Flavell RA: Recognition of double-stranded RNA and activation of NF-kappaB by Toll-like receptor 3. Nature 2001;413:732–738.

40 Hayashi F, Smith KD, Ozinsky A, et al: The innate immune response to bacterial flagellin is mediated by Toll-like receptor 5. Nature 2001;410:1099–1103.

41 Hemmi H, Kaisho T, Takeuchi O, et al: Small antiviral compounds activate immune cells via the TLR7 MyD88-dependent signaling pathway. Nat Immunol 2002;3:196–200.

42 Jurk M, Heil F, Vollmer J, et al: Human TLR7 or TLR8 independently confer responsiveness to the antiviral compound R-848. Nat Immunol 2002;3:499.

43 Bauer S, Kirschning CJ, Hacker H, et al: Human TLR9 confers responsiveness to bacterial DNA via species-specific CpG motif recognition. Proc Natl Acad Sci U S A 2001;98:9237–9242.

44 Hemmi H, Takeuchi O, Kawai T, et al: A Toll-like receptor recognizes bacterial DNA. Nature 2000;408:740–745.

45 Arbibe L, Mira JP, Teusch N, et al: Toll-like receptor 2-mediated NF-kappa B activation requires a Rac1-dependent pathway. Nat Immunol 2000;1:533–540.

46 Benakanakere MR, Zhao J, Galicia JC, Martin M, Kinane DF: Sphingosine kinase-1 is required for toll mediated beta-defensin 2 induction in human oral keratinocytes. PLoS One 2010;5:e11512.

47 Zhao J, Benakanakere MR, Hosur KB, Galicia JC, Martin M, Kinane DF: Mammalian target of rapamycin (mTOR) regulates TLR3 induced cytokines in human oral keratinocytes. Mol Immunol 2010;48:294–304.

48 Brown J, Wang H, Hajishengallis GN, Martin M: TLR-signaling networks: an integration of adaptor molecules, kinases, and cross-talk. J Dent Res 2011;90:417–427.

49 Kanneganti TD, Lamkanfi M, Kim YG, et al: Pannexin-1-mediated recognition of bacterial molecules activates the cryopyrin inflammasome independent of Toll-like receptor signaling. Immunity 2007;26:433–443.

50 Kanneganti TD, Lamkanfi M, Nunez G: Intracellular NOD-like receptors in host defense and disease. Immunity 2007;27:549–559.

51 Lamkanfi M, Kanneganti TD, Franchi L, Nunez G: Caspase-1 inflammasomes in infection and inflammation. J Leukoc Biol 2007;82:220–225.

52 Chamaillard M, Girardin SE, Viala J, Philpott DJ: Nods, Nalps and Naip: intracellular regulators of bacterial-induced inflammation. Cell Microbiol 2003;5:581–592.

53 Chamaillard M, Hashimoto M, Horie Y, et al: An essential role for NOD1 in host recognition of bacterial peptidoglycan containing diaminopimelic acid. Nat Immunol 2003;4:702–707.

54 Girardin SE, Boneca IG, Viala J, et al: Nod2 is a general sensor of peptidoglycan through muramyl dipeptide (MDP) detection. J Biol Chem 2003;278:8869–8872.

55 Mariathasan S, Newton K, Monack DM, et al: Differential activation of the inflammasome by caspase-1 adaptors ASC and Ipaf. Nature 2004;430:213–218.

56 Mariathasan S, Weiss DS, Newton K, et al: Cryopyrin activates the inflammasome in response to toxins and ATP. Nature 2006;440:228–232.

57 Bostanci N, Emingil G, Saygan B, et al: Expression and regulation of the NALP3 inflammasome complex in periodontal diseases. Clin Exp Immunol 2009;157:415–422.

58 Yilmaz O, Sater AA, Yao L, Koutouzis T, Pettengill M, Ojcius DM: ATP-dependent activation of an inflammasome in primary gingival epithelial cells infected by *Porphyromonas gingivalis*. Cell Microbiol 2010;12:188–198.

59 Agostini L, Martinon F, Burns K, McDermott MF, Hawkins PN, Tschopp: NALP3 forms an IL-1beta-processing inflammasome with increased activity in Muckle-Wells autoinflammatory disorder. Immunity 2004;20:319–325.

60 Kato H, Sato S, Yoneyama M, et al: Cell type-specific involvement of RIG-I in antiviral response. Immunity 2005;23:19–28.

61 Hajishengallis G, Lambris JD: Crosstalk pathways between Toll-like receptors and the complement system. Trends Immunol 2010;31:154–163.

62 Krauss JL, Potempa J, Lambris JD, Hajishengallis G: Complementary Tolls in the periodontium: how periodontal bacteria modify complement and Toll-like receptor responses to prevail in the host. Periodontol 2000 2010;52:141–162.

63 Wang M, Krauss JL, Domon H, et al. Microbial hijacking of complement-Toll-like receptor crosstalk. Sci Signal 2010;3:ra11.

64 Hajishengallis G, Wang M, Liang S, Triantafilou M, Triantafilou K: Pathogen induction of CXCR4/TLR2 cross-talk impairs host defense function. Proc Natl Acad Sci U S A 2008;105:13532–13537.

65 Puneet P, Yap CT, Wong L, et al: SphK1 regulates proinflammatory responses associated with endotoxin and polymicrobial sepsis. Science 2010;328 1290–1294.
66 Singh PK, Schaefer AL, Parsek MR, Moninger TO, Welsh MJ, Greenberg EP: Quorum-sensing signals indicate that cystic fibrosis lungs are infected with bacterial biofilms. Nature 2000;407:762–764.
67 Benn M, Hagelskjaer LH, Tvede M: Infective endocarditis, 1984 through 1993: a clinical and microbiological survey. J Intern Med 1997;242:15–22.
68 Marrie TJ, Costerton JW: Morphology of bacterial attachment to cardiac pacemaker leads and power packs. J Clin Microbiol 1984;19:911–914.
69 Marrie TJ, Costerton JW: Scanning and transmission electron microscopy of in situ bacterial colonization of intravenous and intraarterial catheters. J Clin Microbiol 1984;19:687–693.
70 Morris NS, Stickler DJ, McLean RJ: The development of bacterial biofilms on indwelling urethral catheters. World J Urol 1999;17:345–350.
71 Hyde JA, Darouiche RO, Costerton JW: Strategies for prophylaxis against prosthetic valve endocarditis: a review article. J Heart Valve Dis 1998;7:316–326.
72 Gristina AG, Shibata Y, Giridhar G, Kreger A, Myrvik QN: The glycocalyx, biofilm, microbes, and resistant infection. Semin Arthroplasty 1994;5:160–170.
73 Leid JG, Costerton JW, Shirtliff ME, Gilmore MS, Engelbert M: Immunology of staphylococcal biofilm infections in the eye: new tools to study biofilm endophthalmitis. DNA Cell Biol 2002;21:405–413.
74 Leid JG, Shirtliff ME, Costerton JW, Stoodley P: Human leukocytes adhere to, penetrate, and respond to *Staphylococcus aureus* biofilms. Infect Immun 2002;70:6339–6345.
75 Jesaitis AJ, Franklin MJ, Berglund D, et al: Compromised host defense on *Pseudomonas aeruginosa* biofilms: characterization of neutrophil and biofilm interactions. J Immunol 2003;171:4329–4339.
76 Bjarnsholt T, Jensen PO, Burmolle M, et al: *Pseudomonas aeruginosa* tolerance to tobramycin, hydrogen peroxide and polymorphonuclear leukocytes is quorum-sensing dependent. Microbiology 2005;151:373–383.
77 Jensen PO, Bjarnsholt T, Phipps R, et al: Rapid necrotic killing of polymorphonuclear leukocytes is caused by quorum-sensing-controlled production of rhamnolipid by *Pseudomonas aeruginosa*. Microbiology 2007;153:1329–1338.
78 Kesavalu L, Sathishkumar S, Bakthavatchalu V, et al: Rat model of polymicrobial infection, immunity, and alveolar bone resorption in periodontal disease. Infect Immun 2007;75:1704–1712.
79 Guggenheim B, Gmur R, Galicia JC, et al: In vitro modeling of host-parasite interactions: the 'subgingival' biofilm challenge of primary human epithelial cells. BMC Microbiol 2009;9:280.
80 Christersson LA, Zambon JJ, Genco RJ: Dental bacterial plaques: nature and role in periodontal disease. J Clin Periodontol 1991;18:441–446.
81 Marsh PD: Are dental diseases examples of ecological catastrophes? Microbiology 2003;149:279–294.
82 O'Toole G, Kaplan HB, Kolter R: Biofilm formation as microbial development. Annu Rev Microbiol 2000;54:49–79.
83 Davies D: Understanding biofilm resistance to antibacterial agents. Nat Rev Drug Discov 2003;2:114–122.
84 Costerton JW, Stewart PS, Greenberg EP: Bacterial biofilms: a common cause of persistent infections. Science 1999;284:1318–1322.
85 Williams GJ, Stickler DJ: Effect of triclosan on the formation of crystalline biofilms by mixed communities of urinary tract pathogens on urinary catheters. J Med Microbiol 2008;57:1135–1140.
86 Oscarsson J, Karched M, Thay B, Chen C, Asikainen S: Proinflammatory effect in whole blood by free soluble bacterial components released from planktonic and biofilm cells. BMC Microbiol 2008;8:206.
87 Daep CA, Lamont RJ, Demuth DR: Interaction of *Porphyromonas gingivalis* with oral streptococci requires a motif that resembles the eukaryotic nuclear receptor box protein-protein interaction domain. Infect Immun 2008;76:3273–3280.
88 Ramsey MM, Whiteley M: Polymicrobial interactions stimulate resistance to host innate immunity through metabolite perception. Proc Natl Acad Sci U S A 2009;106:1578–1583.
89 Socransky SS, Haffajee AD, Cugini MA, Smith C, Kent RL Jr: Microbial complexes in subgingival plaque. J Clin Periodontol 1998;25:134–144.
90 Tanner A, Maiden MF, Macuch PJ, Murray LL, Kent RL Jr: Microbiota of health, gingivitis, and initial periodontitis. J Clin Periodontol 1998;25:85–98.
91 Goodson JM: Gingival crevice fluid flow. Periodontol 2000 2003;31:43–54.
92 Eskan MA, Benakanakere MR, Rose BG, et al: Interleukin-1beta modulates proinflammatory cytokine production in human epithelial cells. Infect Immun 2008;76:2080–2089.
93 Stathopoulou PG, Benakanakere MR, Galicia JC, Kinane DF: The host cytokine response to *Porphyromonas gingivalis* is modified by gingipains. Oral Microbiol Immunol 2009;24:11–17.

94 Fletcher J, Reddi K, Poole S, et al: Interactions between periodontopathogenic bacteria and cytokines. J Periodontal Res 1997;32:200–205.
95 Steffen MJ, Holt SC, Ebersole JL: Porphyromonas gingivalis induction of mediator and cytokine secretion by human gingival fibroblasts. Oral Microbiol Immunol 2000;15:172–180.
96 Jin L, Soder B, Corbet EF: Interleukin-8 and granulocyte elastase in gingival crevicular fluid in relation to periodontopathogens in untreated adult periodontitis. J Periodontol 2000;71:929–939.
97 Mathur A, Michalowicz B, Castillo M, Aeppli D: Interleukin-1 alpha, interleukin-8 and interferon-alpha levels in gingival crevicular fluid. J Periodontal Res 1996;31:489–495.
98 Tonetti MS, Cortellini D, Lang NP: In situ detection of apoptosis at sites of chronic bacterially induced inflammation in human gingiva. Infect Immun 1998;66:5190–5195.
99 Koulouri O, Lappin DF, Radvar M, Kinane DF: Cell division, synthetic capacity and apoptosis in periodontal lesions analysed by in situ hybridisation and immunohistochemistry. J Clin Periodontol 1999;26:552–559.
100 Iwasaki A, Medzhitov R: Regulation of adaptive immunity by the innate immune system. Science 2010;327:291–295.
101 Ichinohe T, Lee HK, Ogura Y, Flavell R, Iwasaki A: Inflammasome recognition of influenza virus is essential for adaptive immune responses. J Exp Med 2009;206:79–87.
102 Weighardt H, Holzmann B: Role of Toll-like receptor responses for sepsis pathogenesis. Immunobiology 2007;212:715–722.
103 Rodrigues PH, Belanger M, Dunn W Jr, Progulske-Fox A: Porphyromonas gingivalis and the autophagic pathway: an innate immune interaction? Front Biosci 2008;13:178–187.
104 Hajishengallis G, Wang M, Liang S: Induction of distinct TLR2-mediated proinflammatory and proadhesive signaling pathways in response to *Porphyromonas gingivalis* fimbriae. J Immunol 2009; 182:6690–6696.
105 Benakanakere MR, Li Q, Eskan MA, et al: Modulation of TLR2 protein expression by miR-105 in human oral keratinocytes. J Biol Chem 2009; 284:23107–23115.
106 Schmid D, Munz C: Innate and adaptive immunity through autophagy. Immunity 2007;27:11–21.
107 Schmid D, Pypaert M, Munz C: Antigen-loading compartments for major histocompatibility complex class II molecules continuously receive input from autophagosomes. Immunity 2007;26:79–92.
108 Iwasaki A, Medzhitov R: Toll-like receptor control of the adaptive immune responses. Nat Immunol 2004;5:987–995.
109 Kumar H, Koyama S, Ishii KJ, Kawai T, Akira S: Cutting edge: cooperation of IPS-1- and TRIF-dependent pathways in poly IC-enhanced antibody production and cytotoxic T cell responses. J Immunol 2008;180:683–687.
110 Longhi MP, Trumpfheller C, Idoyaga J, et al: Dendritic cells require a systemic type I interferon response to mature and induce CD4+ Th1 immunity with poly IC as adjuvant. J Exp Med 2009;206: 1589–1602.
111 Perrot I, Deauvieau F, Massacrier C, et al: TLR3 and Rig-like receptor on myeloid dendritic cells and Rig-like receptor on human NK cells are both mandatory for production of IFN-gamma in response to double-stranded RNA. J Immunol 2010;185:2080–2088.
112 Beaven MA: Division of labor: specialization of sphingosine kinases in mast cells. Immunity 2007; 26:271–273.

Manju Benakanakere
Department of Periodontics, School of Dental Medicine, University of Pennsylvania
240 South 40th Street
Philadelphia, PA 19104 (USA)
Tel. +1 215 746 4189, E-Mail bmanju@exchange.upenn.edu

Kinane DF, Mombelli A (eds): Periodontal Disease. Front Oral Biol. Basel, Karger, 2012, vol 15, pp 56–83

Neutrophils in Periodontal Inflammation

David A. Scott[a,b] · Jennifer L. Krauss[a]

[a]Center for Oral Health and Systemic Disease and [b]Department of Microbiology and Immunology, University of Louisville, Louisville, Ky., USA

Abstract

Neutrophils (also called polymorphonuclear leukocytes) are the most abundant leukocytes whose primary purpose as anti-microbial professional phagocytes is to kill extracellular pathogens. Neutrophils and macrophages are phagocytic cell types that along with other cells effectively link the innate and adaptive arms of the immune response, and help promote inflammatory resolution and tissue healing. Found extensively within the gingival crevice and epithelium, neutrophils are considered the key protective cell type in the periodontal tissues. Histopathology of periodontal lesions indicates that neutrophils form a 'wall' between the junctional epithelium and the pathogen-rich dental plaque which functions as a robust anti-microbial secretory structure and as a unified phagocytic apparatus. However, neutrophil protection is not without cost and is always considered a two-edged sword in that overactivity of neutrophils can cause tissue damage and prolong the extent and severity of inflammatory periodontal diseases. This review will cover the innate and inflammatory functions of neutrophils, and describe the importance and utility of neutrophils to the host response and the integrity of the periodontium in health and disease.

Neutrophils, the most abundant of the leukocytes, are myeloid-derived anti-microbial professional phagocytes that can also kill pathogens extracellularly, link the innate and adaptive arms of the immune response, and help promote inflammatory resolution and tissue healing. Found extensively within the gingival crevice and epithelium, neutrophils are considered the key protective cell type in the periodontal tissues [1–4]. Intriguingly, histological evidence (fig. 1) suggests that neutrophils form a 'wall' between the junctional epithelium and the pathogen-rich dental plaque [2, 3]. Providing more than a mere barrier function, this anti-microbial wall is thought to function in two ways: as a robust secretory structure (reactive oxygen species, ROS, and bacteriocidal proteins) and as a unified phagocytic apparatus. In other words, the protective power lies in the synergic structure. However, this protection is not without cost, as considerable observational, genetic and experimental

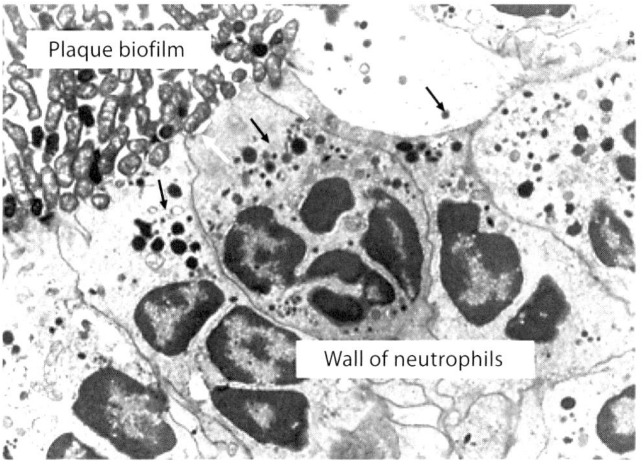

Fig. 1. Interface of bacterial plaque and crevicular neutrophils within the periodontal pocket. Neutrophils in the periodontal pocket forming a wall against the plaque biofilm. Neutrophils cannot engulf the large biofilm structure in vivo. The formation of a wall against the biofilm may be a protective mechanism. Nevertheless, an attempt is made to engulf the surface layer of this biofilm (white arrows). During this process of frustrated phagocytosis, enzymes within neutrophil lysosomal granules (black arrows), products of the oxidative burst, and other proinflammatory substances may be released directly into the pocket and/or the underlying tissue, where they have a predominantly destructive effect. Figure and text reproduced (with permission) from Ryder [3].

data have established a clear association between neutrophil infiltration into the periodontal tissues and the severity and progression of inflammatory periodontal diseases [3, 5–7].

Gene Activity in Neutrophils

While in the bone marrow, neutrophils are transcriptionally and translationally active. During this 'differentiation program', neutrophils are furnished with a large armory of pre-formed signaling and anti-microbial molecules, stored in granules [8]. In the region of $5–10 \times 10^{10}$ new neutrophils are produced daily [9]. Differentiation occurs over approximately 14 days, after which neutrophils enter the circulation as mature, essentially, terminally differentiated cells that have been traditionally considered to be transcriptionally quiescent. However, recent evidence shows that upon leaving the vasculature for infected or inflamed tissues, such as the periodontium, neutrophils undergo a second burst of transcriptional activity, in a process referred to as the 'immune response program' [8]. During the immune response program, cytokines and chemokines, molecules that direct the resolution of inflammation and promote healing, are the prominent translational products [8]. While the relevance

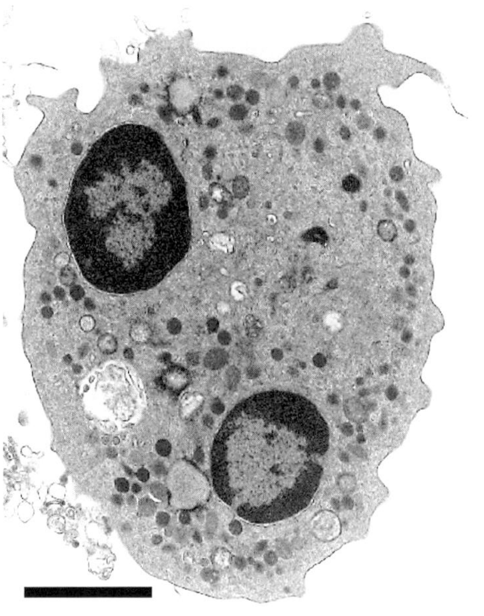

Fig. 2. TEM image of a typical human neutrophil showing intracellular membrane-bound compartments. The multi-lobed nucleus and granule-dense cytoplasm are clearly apparent. Image taken from Guzik et al. [14]. Scale bar = 2 µm.

of neutrophil gene activity in the periodontal tissues has yet to be established, it is important to note that the immune response program renders periodontal neutrophils capable of de novo synthesis of multiple factors that may influence disease progression. In other words, periodontal neutrophils are not functionally reliant solely on their granule contents, as previously thought.

Neutrophil Granule Diversity

A transmission electron microscopy (TEM) image highlighting a typical highly granular human neutrophil is presented in figure 2. The variant membrane-bound intracellular granular structures of neutrophils – known as primary (azurophilic), secondary (specific) and tertiary (gelatinase) granules, as well as the secretory vesicles – are traditionally distinguished by granule-specific biomarker proteins (fig. 3). However, a huge degree of heterogeneity in neutrophil granule content is now appreciated [8]. Granule formation occurs as a continuum, with the considerable heterogeneity explained by the 'targeting-by-timing hypothesis'. This states that different granule proteins are expressed at different stages of the differentiation program, and that these temporally controlled granule proteins are diverted from the constitutive secretory pathway and packaged into granules as and when they are made [8, 10]. Thus, the contents of granules packaged late in the differentiation program will be very different than those packaged early in the differentiation program. The ability

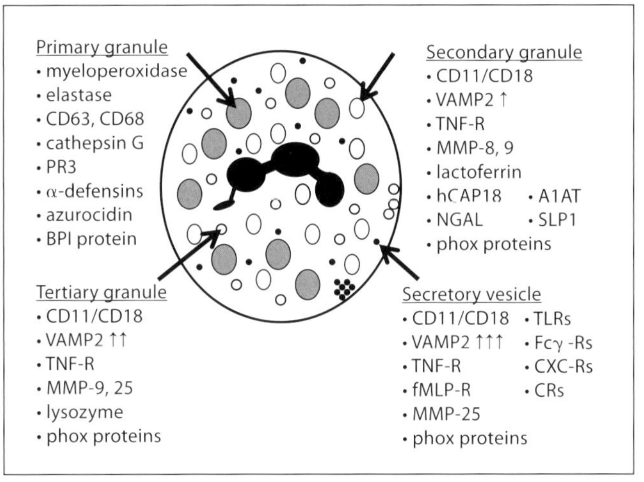

Fig. 3. Contents of the intracellular membrane-bound compartments of human neutrophils. CD63 and CD68 are degranulation markers. VAMP2 is key in granule membrane-cell membrane fusion and exocytosis. Phox proteins are NADPH oxidase subunits involved in ROS production (Gp91phox; p22phox). PR3 = Proteinase 3; TLRs = TLR-1, 2, 4, 6 and 8; CXC-Rs = CXC receptors 1, 2 and 4 as well as chemokine receptors 1, 2 and 3; CRs = complement receptor 1 (CD35), c1qR. Data presented are combined from various reports [8, 12, 13, 15].

of the differing granule sub-types to fuse with the neutrophil cell membrane and, thus, degranulate into the extracellular environment or populate the neutrophil cell surface (as opposed to fusing with the phagosomal membrane) is similarly temporally controlled. v-SNAREs [vesicle-soluble NSF (N-ethylmaleimide sensitive factor/fusion protein) attachment protein receptors] are produced in larger amounts as the differentiation program progresses, with vesicle-associated membrane protein-2 (VAMP-2) the best characterized [8, 11]. v-SNAREs, such as VAMP-2, enable granule membrane fusion with the cell membrane. Thus, the targeting-by-timing hypothesis also provides an explanation for the alternate functions of the various neutrophil granule compartments. The last granules to form in the bone marrow – the secretory vesicles – contain the greatest concentration of v-SNAREs, and thus they control neutrophil environmental responsiveness and firm adhesion to activated vascular endothelia in the periodontium and elsewhere. The tertiary granules facilitate extravasation via matrix metalloproteinase (MMP)-mediated degradation of the basement membrane (without release of the degradative neutrophil serine proteases, i.e. cathepsin G, neutrophil elastase and proteinase 3). The secondary granules promote phagocytotic capacity, while primary and secondary granules each contribute the major anti-microbial arsenal [8, 12, 13]. While not an absolute, then, the predominant proteinaceous contents of the various intracellular membrane-bound compartments in human neutrophils are presented in figure 3. The combination of proteases shown in figure 3 (in conjunction with ROS) is capable of destroying most components of the periodontal soft tissues, not limited to collagen and collagen-derived peptides, including the organic component of alveolar bone, fibrinogen, fibronectin, laminin and elastin.

As is clear from figure 3, multiple receptors are stored in the v-SNARE-rich secretory vesicles and can be readily transported to the cell surface. The interaction of selectin adhesion molecules present on the neutrophil surface with activated vascular endothelium, as is present in abundance during periodontal inflammation, is sufficient to induce the intracellular calcium fluxes required to induce secretory vesicle fusion with the cell membrane [8, 16]. Human neutrophils express all known TLRs (TLR1–10), with the exception of TLR3, with v-SNAREs enabling the transfer of TLRs that signal from the surface (TLR1, 2, 4 and 5) [17, 18]. The importance of TLRs in recognizing periodontal and other pathogens by innate cells is addressed in detail in the chapter by Benakanakere and Kinane [pp. 41–55]. Other signaling molecules transposed to the neutrophil surface include extrinsic apoptosis receptors (e.g. FAS, TNFRI, TRAIL-R2), apoptosis inhibitors (IL-1R, TLR4) and anti-apoptosis receptors (e.g. for IFNγ and GM-CSF). While this array of environmental sensing receptors are present intracellularly prior to exiting the bone marrow, it is not until inflammatory stimuli are encountered that these signaling molecules are transported to the cell surface, rendering them functional.

Bacteria-Killing Mechanisms of Neutrophils

Until the recent growth in our understanding of the important roles played by epithelial cells in innate immunity (see Benakanakere and Kinane [pp. 41–55]), neutrophils have been considered the first line of defense against infectious agents. Neutrophils can destroy periodontal pathogens by both oxygen-dependent (the 'respiratory or oxidative burst') and oxygen-independent (lytic and proteolytic enzymes) mechanisms.

Functional neutrophil NADPH oxidase, required for ROS generation, is a composite of 7 or more proteins. Cytosolic NADPH oxidase components are translocated to the phagosome membrane upon 'neutrophil priming' (engagement by one or more proinflammatory messengers), where they assemble into NADPH oxidase complexes in association with phagosomal cytochrome *b*558 [17, 19]. NADPH oxidase catalyzes the production of superoxide anions. Granule proteins also contribute to the production of ROS within phagosomes, most notably myeloperoxidase, which produces $OH^{\bullet}$ and singlet oxygen as well as hypochlorous acid from H_2O_2 and chloride [17, 19]. The strong oxidant combination generated by the respiratory burst and myeloperoxidase is sufficient to kill most prokaryotes through compromise of bacterial phospholipid bilayers, fragmentation/inactivation of proteins and induction of DNA damage.

Several granule proteins are involved in oxygen-independent bacterial killing. In hypoxic periodontal pockets this may be critical, although neutrophils may be capable of generating ROS in periodontal pockets with oxygen concentrations as low as 1–3% [20]. Indeed, the generation of neutrophil serine protease-deficient mice has evidenced the importance of cathepsin G and elastase in the destruction of both Gram-positive and Gram-negative microbes [21, 22]. Azurocidin (hCAP 37kDa), cathelicidin (LL-37), human neutrophil peptides (HNPs; α-defensins), lactoferrin, elastase and lysozyme are all neutrophil-derived molecules with direct or indirect antimicrobial activities [23, 24]. The most common mechanism of neutrophil-derived antimicrobial action is disruption of the integrity of the bacterial cell membrane (cathelicidin; α-defensins; serine protease; azurocidin). Other granule-derived antimicrobial mechanisms include destruction of peptidoglycan (lysozyme), iron sequestration (lactoferrin; neutrophil gelatinase-associated lipocalin), opsonization (azurocidin) and the degradation of proteolytic bacterial virulence factors (elastase) [8, 22, 24]. For a more complete review of neutrophil granule-dependent killing mechanisms, please see Soehnlein [24].

While β-defensins are predominantly produced by the oral epithelia, neutrophils produce mainly α-defensins (HNP1–4) [25]. Increased concentrations of α-defensins have been shown in neutrophils in diseased periodontal tissues [25]. It seems that periodontal pathogens (e.g. *P. gingivalis*, *Treponema denticola*) in general may be more resistant to neutrophil phagocytosis and to epithelial (human β-defensin-3)-derived and neutrophil (cathelicidin)-derived antimicrobials than non-pathogens (e.g. *Actinomyces naeslundii* and *Streptococcus sanguis*) [26]. Cathelicidin, for example, can neutralize the anti-neutrophilic leukotoxin produced by *Aggregatibacter actinomycetemcomitans* [27]. However, it seems that neutrophil phagocytosis of *A. actinomycetemcomitans* may be inefficient regardless of strain or ability to produce leukotoxin [28].

This plethora of oxygen-dependent and oxygen-independent antimicrobial factors – in addition to extracellular bacterial traps (see below) – renders neutrophils very efficient at controlling periodontal pathogens and, generally, results in containment of pathogens within the oral cavity. The importance of functional neutrophils in the maintenance of periodontal health will be discussed later in the context of genetic neutrophil defects that predispose carriers to often severe inflammatory periodontal diseases. Conversely, the contribution of overly robust or prolonged neutrophilic anti-microbial activities to collateral periodontal tissue damage will also be discussed.

Delivery of Neutrophils to the Periodontal Tissues

Neutrophil recruitment requires adhesion to, and transmigration through, blood vessel walls at sites where the vascular endothelium is activated by proinflammatory mediators. The human junctional epithelium is never sterile, meaning that even with optimal plaque control, neutrophils will still be stimulated to exit the gingival microvasculature, enter the periodontal tissues and, subsequently, migrate firstly toward endogenous epithelium-derived (such as IL-8 and IL-1β) and serum-derived (plaque activated C5a) chemoattractants then preferentially toward exogenous chemotactic signals (such as lipopolysaccharides, LPS, and formylated peptides like f-methionine-leucinephenylalanine, fMLP) produced by plaque bacteria in the gingival crevice. Indeed, high levels of bacteria and individual neutrophil chemoattractants, including stromal-cell-derived factor-1 (SDF-1α or CXCL12) [29], are found in the gingival crevice, which can result in the recruitment of vast numbers of neutrophils into the periodontal pocket [30]. Neutrophils can employ glycolysis to meet their energy requirements, whereas macrophages must rely on oxidative phosphorylation. Hence, neutrophils are the key phagocytic defenders in the periodontal pocket. Essentially, the stronger the inflammatory stimulus, the greater the epithelial and endothelial activation and the larger the number of neutrophils recruited.

Some oral pathogens produce virulence factors that inhibit neutrophil transmigration into the periodontal tissues, promote leukocyte longevity or are cytotoxic to recruited neutrophils. The HAD family serine phosphatase, SerB, produced by *P. gingivalis*, for example, may play an important role in preventing granulocyte recruitment to the periodontal tissues as, in a rat model, there is greater neutrophil infiltration upon infection with an isogenic ΔSerB mutant compared to wild-type bacteria [31]. On the other hand, neutrophils treated with supernatants from monocytes that have been exposed to *P. gingivalis* LPS exhibit suppressed apoptosis relative to supernatants from naïve monocytes, suggesting a longer functional life for neutrophils [32]. The best characterized plaque-derived anti-neutrophil molecule is the leukotoxin LtxA, produced by *A. actinomycetemcomitans* [33]. LtxA is a member of the RTX (repeat in toxin) group of toxins that targets αL/β2 (LFA-1; Cd11b/CD18) adhesion molecules on the surface of hematopoietic cells and is highly cytolethal to neutrophils through induction of apoptotic cell death and/or formation of pores in the innate cell membrane [33, 34].

Neutrophil Extracellular Traps in Periodontal Pockets

Neutrophil extracellular traps (NETs) are extracellular neutrophil-derived chromatin fibers combined with various granule proteins (including: elastase; cathepsin G; myeloperoxidase; bactericidal/permeability-inducing protein, BPI; lactoferrin; peptidoglycan recognition proteins; MMP-9) that bind and kill bacteria, and other pathogens, as well as destroying microbial virulence factors, presumably proteolytically [35, 36].

Several oral pathogens have long been known to produce factors that can subvert the immune response. The discovery of NETs provides a mechanism by which innate cells can 'fight back' or, in other words, interfere with bacterial virulence factors extracellularly. NETs may also control innate cell life span through an alternate mechanism of eukaryotic cell death, known as NETosis. An abundance of NETs containing trapped bacteria can be visualized in periodontal pockets (pocket surface; gingival crevicular fluid, GCF; and pus) [37–39], as shown in figure 4. The study of NETs in altering pathogen-host interactions is in its infancy and the role of NETs in protecting against periodontal diseases has yet to be established. However, NETs are clearly present in the periodontal crevice and, furthermore, it has been hypothesized that bacteria that produce DNAses capable of degrading the chromatin NET backbone may be more virulent than those without DNAse [35, 40]. Additionally, DNAse-producing bacteria may facilitate the maturation of a pathogenic plaque biofilm. It is hoped that the next few years will provide fascinating insights in the role played by NETs in the maintenance of periodontal health and how periodontal pathogens may have evolved strategies to subvert NET-mediated bacterial killing.

Neutrophil Survival in the Periodontal Tissues

On leaving the bone marrow, neutrophils generally remain in circulation for less than 12 h. Circulating neutrophils that do not extravasate into inflammatory compartments are catabolized in the spleen [1]. Neutrophils that do enter tissues undergo apoptosis after 1–2 days and are subsequently cleared by macrophages. Conventional extrinsic (e.g. TNF-, FAS ligand- and TRAIL-induced; caspase-dependent and independent) and intrinsic (spontaneous, constitutive and, e.g., ROS-induced or mitochondria-induced in senescent cells) apoptotic pathways are clearly operational, as recently reviewed [17]. However, other less well-defined cell death programs are also active in neutrophils, including phagocytosis-induced cell death and NADPH-oxidase-dependent NETosis, while neutrophils also appear to have a large autophagic capacity [17, 41–43]. It is important to note that the de novo production of proinflammatory cytokines is suppressed, and most primary effector functions are downregulated, during apoptosis [17, 44, 45]. During apoptosis multiple redundant macrophage receptors facilitate effective recognition, engulfment and clearance of neutrophils. Such pro-phagocytotic molecules include complement receptors, the phosphatidylserine receptor and several other scavenger receptors, such as MARCO and CD36 [17, 41]. Furthermore, the uptake of apoptotic neutrophils induces the production of anti-inflammatory cytokines, such as TGFβ, in the phagocytizing cell [17, 46]. Several periodontal pathogens have been shown to induce phagocytosis-induced cell death in innate cells, including *Fusobacterium nucleatum* and *P. gingivalis*, as well as other bacteria known to inhabit the oral cavity, such as *Neisseria gonorrhoeae* [17, 47, 48]. Induction of apoptosis presumably benefits periodontal pathogens by preventing phagocytosis. In general, however, apoptosis of inflammatory cells, particularly neutrophils, is key to the resolution of

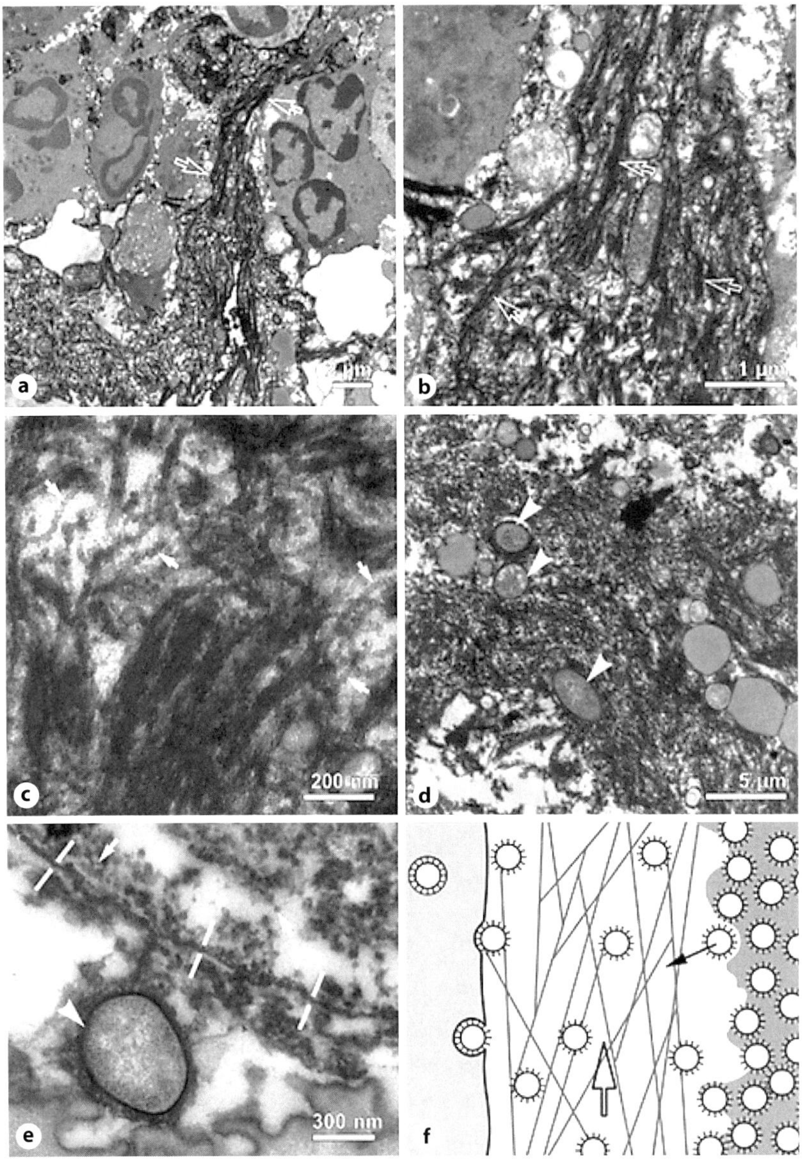

Fig. 4. Extracellular NETs in the periodontitis. TEMs of pocket epithelium biopsies are presented. **a** Overview of neutrophil extracellular traps on the pocket epithelium surface. The pronounced blackening is characteristic for neutrophil extracellular traps. The open arrows indicate neutrophil extracellular traps threads. **b** Detail of **a**. The threads (open arrows) consist of individual fibers that run in parallel. **c** High magnification reveals that neutrophil extracellular traps consist of uniformly thick fibers (small solid arrows) with a diameter of nearly 15–17 nm. **d** Solitary bacteria (indicated by solid white arrowheads) entrapped by crevicular neutrophil extracellular traps. **e** A bacterium (solid white arrowhead) adherent to the epithelium surface is entrapped by neutrophil extracellular traps. A longitudinally running neutrophil extracellular trap fiber is delineated with white bars. Small solid white arrows: a multitude of cross-sectioned neutrophil extracellular trap fibers. **f** A scheme of a crevice. On the left: an epithelial cell with an adherent bacterium and bacteria in different stages of internalization. In the middle: periodontal crevice. Neutrophil extracellular traps build a 'firewall', protecting the gingiva from bacteria dispersing out of the subgingival plaque. The neutrophil extracellular

inflammation and prevention of collateral tissue damage by degradative neutrophil enzymes and ROS that would otherwise be released into the periodontium upon neutrophil necrosis and/or degranulation. In other words, apoptosis is a non-phlogistic process.

On the other hand, several inflammatory mediators promote neutrophil longevity at sites of inflammation – such as LPS, lipoteichoic acids, C5a, IL-1α, GM-CSF, G-CSF and IFNγ – rendering these granulocytes resistant to extrinsic ligand-induced apoptosis (FAS and TNF) [49]. The consequence is extended functionality [17]. Some of the molecules known to extend neutrophil life span also prime neutrophils in a manner that enhances ROS production, and perhaps other functions, when a second activating signal is encountered [17]. Many of these longevity-inducing factors are present in high concentrations in periodontal tissues and gingival crevicular fluid in inflammatory periodontal diseases, particularly LPS produced by the predominantly Gram-negative flora associated with periodontitis. Some pathogens have developed mechanisms for suppression of apoptosis in neutrophils, such as *Chlamydia pneumoniae* [50]. However, the potential antiapoptotic properties of the periodontal microflora have not received much research attention. Specific bacterial pathogens may also promote inflammation through the promotion of degranulation and/or lysis and/or secondary necrosis in neutrophils – most notably *Streptococcus pyogenes* and leukotoxin-producing *A. actinomycetemcomitans* strains [51–53]. While the balance between pro- and antiapoptotic signals for neutrophils in periodontal tissues has not been clearly established – and is likely to vary depending on disease state and environmental influences – there are clearly large numbers of functional neutrophils present in the gingival tissues during gingivitis and at all stages of chronic periodontitis capable of both killing bacterial pathogens and contributing to collateral tissue damage.

Neutrophils and Promotion of Edema

Neutrophils are thought to play key roles in promoting edema as a consequence of interactions with the vascular endothelia during diapedesis through the secretion of arachidonic acid derivatives, chemokines (such as CXCL1, 2, 3 and 8) and heparin-binding protein (CAP37 or neutrophil azurocidin 1), with neutrophil elastase and cathepsin G also reported to promote vascular permeability, at least in vitro [13, 54–56]. Of these mediators, the arachidonic acid metabolites may be of particular importance due to their ability to enhance the permeability-inducing capacities of

traps, together with the entrapped bacteria, are continuously pushed into the oral cavity by the crevicular exudate outflow. Open white arrow: direction of the crevicular exudate outflow. On the right: subgingival plaque. Solid black arrow: a bacterium dispersing from the subgingival plaque. Figure and text reproduced (with permission) from Vitkov et al. [39].

other endogenous (e.g. C5a, histamine) and exogenous (e.g. fMLP) proinflammatory mediators. Several factors generated de novo by neutrophils during the immune response program help to promote edema, most notably TNF, CXCL2 and CXCL8 [13, 57, 58]. Finally, in addition to edema induced by the aforementioned permeability-increasing (and proangiogenic) factors, neutrophils also contribute to edema by damaging the integrity of the vascular endothelium through the actions of ROS and serine proteases [13, 59, 60].

Neutrophil-Mediated Tissue Destruction

It is generally accepted that the inflammatory nature of the neutrophil response to the oral biofilm can promote a homeostatic imbalance that is key to periodontal disease onset and progression. As reviewed in 2010 [3], there are two prominent mechanisms that have been proposed to explain the role of neutrophils in periodontal disease development. These are:

- the impaired neutrophil;
- the hyperactive neutrophil.

A third mechanistic category could also explain the role of neutrophils in periodontal disease development:

- chronic recruitment and activation of the normal neutrophil.

In other words, it could be that destructive periodontal disease, the prevalence of which increases with age, is the price to be paid for the control of periodontal pathogens through the chronic activation and extended longevity of normal neutrophils over many years.

The weight of evidence seems to suggest that overt neutrophil defects generally lead to a predisposition towards aggressive forms of periodontitis [3]. Impairment of neutrophil function(s) in such cases is frequently genetically (or intrinsically) determined, as discussed further below. However, the importance of the impaired or defective phenotype in disease progression is often exacerbated by environmental factors, most notably by cigarette smoking [62].

Hyperactivity, with respect to neutrophils, has been described in the periodontal literature to mean 'elevated function', e.g. increased enzymatic activity, particularly an increased respiratory burst [3]. The term hyperactive is often used synonymously with 'primed', although it is not entirely clear that this is always appropriate. All neutrophils can be primed by various proinflammatory mediators but some authors apply a variant meaning [63, 64]: a specific and intrinsic predisposition to elevated function in neutrophils in those predisposed to inflammatory periodontal diseases. Therefore, there is a need to clearly differentiate between neutrophils that have been activated, or primed, as a consequence of extravasation and neutrophils that are hyperactive prior to activation by local periodontal stimuli. That is, neutrophils that are already primed while in systemic circulation. Again, for an in-depth review of priming versus impairment of function of neutrophils and the consequences for aggressive versus chronic periodontitis, please see the recent review by Ryder [3].

Certainly, normal neutrophils that enter the periodontal environment are primed. This is as a consequence of the adhesive interactions with activated vascular endothelium and endogenous proinflammatory mediators during the extravasation process [65] and/or exposure to intact periodontal bacteria, such as *P. gingivalis* [66] and *A. actinomycetemcomitans* [67], or bacteria-derived proinflammatory mediators, such as fMLP [3, 66].

However, certain gene polymorphisms may lead to the hematopoiesis of neutrophils that exhibit intrinsically elevated function. For example, peripheral, presumably naïve, neutrophils from subjects with a specific [131 H/H] polymorphism in the neutrophil Fcγ receptor IIa (FcγRIIa) express higher levels of the degranulation markers, CD63 and CD66b, and release more elastase on exposure to immune serum-opsonized *A. actinomycetemcomitans* than neutrophils from the more common FcγRIIa genotype (131 R/R) [68]. Thus, 131 H/H FcγRIIa neutrophils are hyperactive without the need for priming. Individuals with 131 H/H FcγRIIa are also predisposed to more severe periodontitis than those with the 131 R/R FcγRIIa genotype [68]. Further, Matthews et al. [69] reported that peripheral neutrophils from subjects with periodontitis produced higher unstimulated levels of extracellular ROS than granulocytes from control subjects, and that this phenomenon was unaffected by priming by *P. gingivalis* and *F. nucleatum*. On the other hand, effective periodontal therapy reduced FcγR-stimulated ROS production. Multiple studies have confirmed that peripheral neutrophils isolated from chronic periodontitis patients, compared to control neutrophils, produce higher levels of ROS on exposure to neutrophil-activating signals, including LPS and whole bacteria [70, 71].

Even when the major endogenous mediators of periodontal tissue destruction, the serine proteases and matrix metalloproteinases [72], are released extracellularly via necrosis, exocytosis/degranulation or via NETs, there are multiple endogenous inhibitors that limit their activity. These include α1-proteinase inhibitor (neutrophil elastase and proteinase 3), elafin (neutrophil elastase and proteinase 3), secretory leukocyte protease inhibitor (cathepsin G and neutrophil elastase), α1-chymotrypsin (cathepsin G) and the tissue inhibitors of metalloproteinases (TIMPs) [22, 73]. However, it is thought that proteolytic enzymes that are either surface-associated or present in tight granulocyte-substrate junctions may be protected from endogenous inhibitors and that high local protease gradients – as occur in diseased periodontal tissues – can overwhelm the endogenous inhibitory capacity [22]. Furthermore, unlike most MMP-producing cells, neutrophils do not contain or manufacture de novo the major TIMP, TIMP-1 [73]. Total proteolytic activity [74] as well as multiple neutrophil-derived proteolytic enzymes have been shown to be elevated in periodontitis, compared to healthy controls, including elastase (GCF, periodontal tissues) [75–78], proteinase 3 (GCF cell pellet) [74], cathelicidin (GCF and periodontal tissues) [25, 79–81], myeloperoxidase

(GCF) [82] and matrix metalloproteinases whose major source in the periodontium is the neutrophil (MMP-2, MMP-8, MMP-9 and MMP-25) [75, 83–88]. To this end, it is clear that exposure to (and/or phagocytosis of) several periodontal pathogens – including *F. nucleatum*, *T. denticola*, *P. gingivalis* and *A. actinomycetemcomitans* – leads to the extracellular release of tissue-degrading molecules, such as ROS, MMP-8, MMP-9 and elastase from neutrophils, as well as proinflammatory cytokines, e.g. IL-1β, that are expected to prolong the inflammatory response [89–91]. A prolonged neutrophil response to plaque may also contribute indirectly to compromised wound healing. For example, elastase is efficient at cleaving platelet-derived growth factor receptors from periodontal ligament cells, compromising proliferative signaling pathways in response to this important growth factor [92].

Interestingly, recent data has suggested a potential oral-systemic inflammatory activation loop for neutrophils. Dias *et al* have shown that plasma from periodontitis subjects is more efficient than plasma from healthy subjects in priming neutrophils to fMLP responsiveness and in directly inducing the oxidative burst. Antibodies against IL-8, GM-CSF and IFN-α abrogated this superoxide-inducing potential of plasma [93]. Others have shown that *F. nucleatum* exposure results in the upregulation of multiple ROS-related genes in neutrophils and that pro-inflammatory transcripts, including several ROS response-related genes, are differentially regulated in cells from periodontitis subjects and healthy controls [94].

In addition to breaking down collagen, and other proteins, to provide a carbon source, the gingipains produced by *P. gingivalis* are capable of modulating multiple factors that influence neutrophil recruitment or function. For example, R- and K-gingipain treatment of the 77 amino acid long variant of IL-8 results in N-terminal cleavage of 5 to 11 amino acids from this chemokine, with the truncated IL-8 molecule exhibiting enhanced chemotactic potency and increased priming activity in neutrophils, as measured by ROS production on subsequent exposure to fMLP. The opposite phenomenon (decreased neutrophil chemotaxis and priming) occurs when the more potent, innate cell-derived 72 amino acid IL-8 variant is treated with gingipains [95]. *T. denticola* produces a surface-associated C3 to iC3b converting protease, dentilisin, that promotes ROS production in human neutrophils [91]. It is likely that multiple host-pathogen interactions influence the release and activity of endogenous mediators of tissue destruction.

Neutrophils Tune the Immune Response

Recent evidence shows that neutrophil-derived proteases also modulate chemokine activity [22]. For example, limited cleavage of IL-8 (CXCL8) by proteinase 3 increases the chemoattractive potency of this key proinflammatory mediator [96], while cathepsin-G-mediated cleavage lowers the chemotactic potency of CCL5 (RANTES) [97]. Similarly, neutrophil-derived proteases can activate some cytokines, most notably

TNF and IL-1β [98] and inactivate others, such as IL-6 [99]. Azurocidin and proteinase-3 upregulate adhesion molecules on the vascular endothelium, while cathelicidin is a potent recruiter of monocytes to sites of bacterial infection [24]. Thus, these granule proteins combine to promote the intensity and longevity of the inflammatory response. Neutrophil proteases may also amplify and link the innate and adaptive arms of the immune system. Chimerin (tazarotene-induced gene 2 protein, TIG-2; retinoic acid receptor responder protein 2) is an important chemoattractant to professional antigen-presenting cells (dendritic cells and macrophages) required to engage lymphocytes. Both cathepsin G and neutrophil elastase can convert prochemerin into the active form, chemerin [100]. Indeed, it seems that neutrophil serine proteases and MMPs can, between them, modulate multiple molecules that play important roles in inflammation and tissue remodeling, including: anti-bacterial peptides (e.g., hCAP18 conversion to cathelicidin; azurocidin is chemotactic for, and an activator of, monocytes/macrophages [101]) [27]; growth factors (e.g., VEGF and TGFβ release); adhesion molecules (e.g., ICAM-1 and VCAM-1 processing); and a host of cell surface signal transducers (e.g., TLR4 activation, TNFR inactivation) [22, 102, 103]. While full discussion is beyond the scope of this article, the activities and consequences of neutrophil-derived protease interaction with multiple biological targets have been recently reviewed elsewhere [22, 102, 103].

Neutrophil Defects and Polymorphisms Associated with Aggressive Periodontitis

The importance of fully functional neutrophils to the maintenance of good periodontal health is dramatically highlighted by the large number of genetic neutrophil defects associated with, often severe, inflammatory periodontal disease. The periodontal characteristics of some well-defined neutrophil defects are listed below[1].

Neutropenias (<1,500 neutrophils per ml of blood) including *Kostmann Syndrome* (neutropenia with reduced cathelicidin and HNP1–3): susceptibility to recurrent bacterial infection and aggressive periodontitis [3, 25, 104–106];

- *Chediak-Higashi syndrome* [LYST (lysosomal trafficking regulator) mutation resulting in impaired phagolysosome formation]: aggressive periodontitis [107–110];
- *Papillon-Lefèvre syndrome/dipeptidyl peptidase I deficiency* (Cathepsin C mutation): aggressive periodontitis [107, 111–113]; periodontal manifestations of Papillon-Lefèvre syndrome in a single patient are shown in figure 5;
- *A1AT-deficiency*: periodontitis [114–116];
- *Leukocyte adhesion deficiency* (CD18 integrin or sialyl-LewisX mutation): recurrent bacterial infections and aggressive periodontitis [107, 117–120];

[1] Aggressive periodontitis includes earlier terminologies such as juvenile periodontitis, early onset periodontitis and rapidly progressive periodontitis. Aggressive periodontitis occurs in 0.1–15% of the population, although valid prevalence estimates are problematic [130].

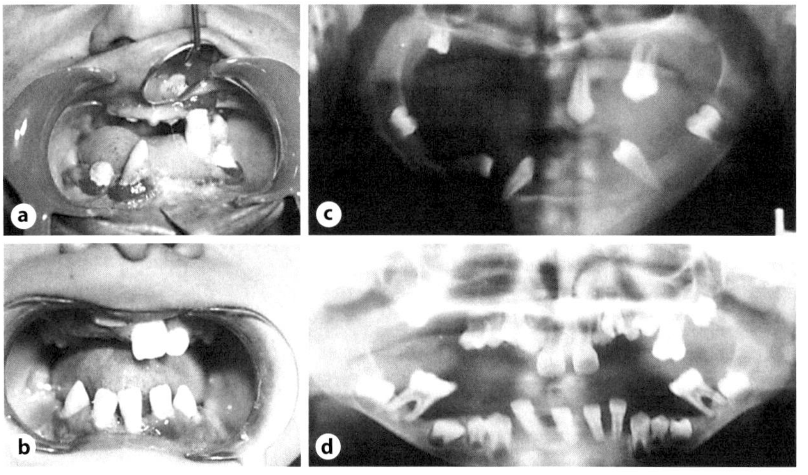

Fig. 5. Periodontal manifestations of Papillon-Lefèvre syndrome. **a**, **b** Photographs showing marked red inflamed and swollen gingivae. **c**, **d** Radiograph showing severe loss of alveolar bone. Images reproduced (with permission) from Dhanrajani [139].

- *Granulomatous disease* (defective NADPH oxidase system): susceptibility to bacterial infection and periodontitis [107, 121, 122];
- *NA2 polymorphism in FcγRIIIb* (a neutrophil-specific antibody receptor): aggressive periodontitis [123, 124];
- *Multiple single nucleotide fMLP-R polymorphisms*[2]: aggressive periodontitis [3, 125–129]

The above list is not intended to be exhaustive and the number of potential identified genetic defects that affect neutrophil function and predispose carriers to periodontitis continues to increase. For example, lysosomal-associated membrane protein-2 (LAMP-2) seems critical in directing appropriate phagosomal maturation. LAMP-2 knockout mice (a model of Danon Disease) exhibit reduced bacteriocidal activity and develop severe periodontitis at an early age [135–136]. Individuals with neutrophil-specific granule deficiency exhibit recurrent bacterial infections [137], but studies of the oral health of such subjects were not identified by the current search strategy. Interestingly, a polymorphism in p22phox NADPH oxidase (C242T) – a key gene in the generation of the respiratory burst in neutrophils – may also be associated with aggressive periodontitis [138].

[2] Reviews of associations between other single nucleotide polymorphisms – including polymorphisms in IL-1β, IL-4, IL-10, TNF, TLR, fMLP receptor and human leukocyte antigen genes – and susceptibility to periodontal diseases are available [1, 131–134].

Neutrophils in Aging Subjects

Immunosenescence occurs as we age and is accompanied by an increased susceptibility to multiple infectious diseases, including periodontal diseases [61]. Even the acute inflammation associated with experimentally-induced gingivitis may be more severe in older subjects [140, 141]. There are limited studies on the effects of age on neutrophil recruitment and function. Of those functions so far examined (expression of key receptors: GMCSF-R, TLR4 and TLR2; apoptosis; hematopoiesis; chemotaxis; phagocytosis; oxidative burst; and specific activating pathways), it seems that chemotaxis, phagocytosis and capacity to signal via calcium and MAPK-dependent pathways may be impaired [reviewed in 61, 142, 143]. Thus age-associated depreciations in neutrophil function may help explain increased susceptibility to periodontitis in aging populations.

Smoking and Neutrophil Function

Genetic and age-induced defects in neutrophil function increase the risk of periodontitis. Similarly, tobacco-induced innate dysfunction appears to contribute to a predisposition towards inflammatory periodontal diseases. Indeed, crevicular neutrophil viability has been shown to be consistently reduced in smokers (approximately 75% viable) compared to non-smokers (approximately 85% viable), with phagocytotic capacity also diminished in a potentially dose-dependent manner (40% in heavy smokers; 79% in non-smokers) [144]. Furthermore, overall transcriptional activity associated with the immune response program is lower in neutrophils isolated from smokers compared to non-smokers [145].

As we have recently reviewed [146], tobacco smoke exposure induces systemic neutrophilia [147] and exerts profound effects on human neutrophils, including compromised neutrophil maturation [147, 148] and inefficient effector function [148–150]. Key to the development of destructive inflammatory periodontal disease may be the tobacco-induced induction of neutrophil-derived elastase [151, 152] and metalloproteinase release [148, 153, 154] combined with compromised phagocytotic and bacteriocidal capacities, including diminished ability to kill *P. gingivalis* [148, 155, 156] and impaired chemotactic responsiveness [150].

Neutrophils express functional receptors for nicotine and cotinine [148, 157], as well as tobacco-derived aryl hydrocarbons [158]. Neutrophils also express receptors for several endogenous immunomodulatory factors that have been shown to be dysregulated in tobacco smokers. These factors include IL-8, ICAM-1 and TNF [150, 159–162]. The α7 nicotinic acetylcholine receptor (nAChR)-triggered cholinergic anti-inflammatory pathway is a key endogenous mechanism by which an overly robust proinflammatory response to infection is prevented [163–165]. Neutrophils express abundant α7-nAChR, as shown in figure 6, from an early stage of differentiation [148]. The role of the cholinergic anti-inflammatory pathway in suppressing

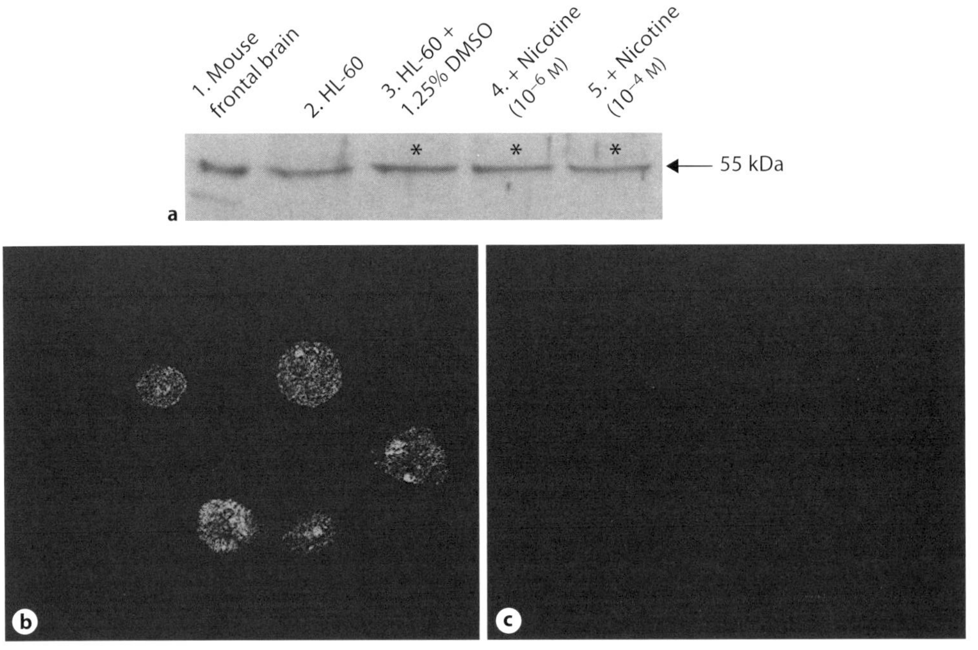

Fig. 6. Promyelocytic and neutrophilic HL-60 cells express α7-nAChR. **a.** α7-nAChR Western blot. Lane 1: mouse frontal brain extract (10 μg; positive control); lane 2: lysate (40 μg) of promyelocytic HL-60 cells; lanes 3–5: lysate (40 μg) of 5-day DMSO-induced (neutrophilic) HL-60 cells without nicotine, or with 10^{-6} M nicotine, or with 10^{-4} M nicotine treatment, respectively. α7-nAChR-specific antibodies (Santa Cruz SC-5544) bound to protein bands exhibiting a relative molecular mass of 55.0 kDa in all lanes. Densitometric analysis (data not shown) revealed that α7-nAChR-specific band intensities were significantly increased in the DMSO-treated samples, relative to promyelocytic cells (n = 5), but that this differentiation-associated increase in α7-nAChR protein level was not influenced by nicotine exposure during differentiation in any statistically significant manner. * $p < 0.05$. **b** α7-nAChR immunofluorescence staining: fixed 5-day DMSO-induced HL-60 cells were incubated with a polyclonal rabbit anti-α7 antibody (Santa Cruz SC-5544) before staining with a FITC-conjugated anti-rabbit antibody (Santa Cruz SC-2253). α7-nAChR-specific staining was evenly distributed across the cells. A similar cellular distribution of α7-nAChR was observed in promyelocytic cells (data not shown). ×1,000. **c** Negative control for α7-nAChR immunofluorescence staining. ×1,000. This figure was originally published by BioMed Central in Xu et al. [148].

proinflammatory cytokine production during the neutrophilic immune response program remains to be clarified. However, clearly, α7-nAChR agonists are capable of reducing the expression level of the key adhesion molecule CD11b and suppressing transalveolar neutrophil migration in animal models of pulmonary inflammation and other inflammatory diseases [166–170]. Furthermore, nicotine-induced suppression of the oxidative burst and *P. gingivalis* killing in neutrophils is probably α7nAChR-dependent, and certainly α-bungarotoxin (an α7-nAChR inhibitor) sensitive [148]. A model of the mechanistic consequences of the inappropriate, i.e. not infection- or trauma-induced acetylcholine driven, engagement of α7-nAChRs on neutrophils is presented in figure 7.

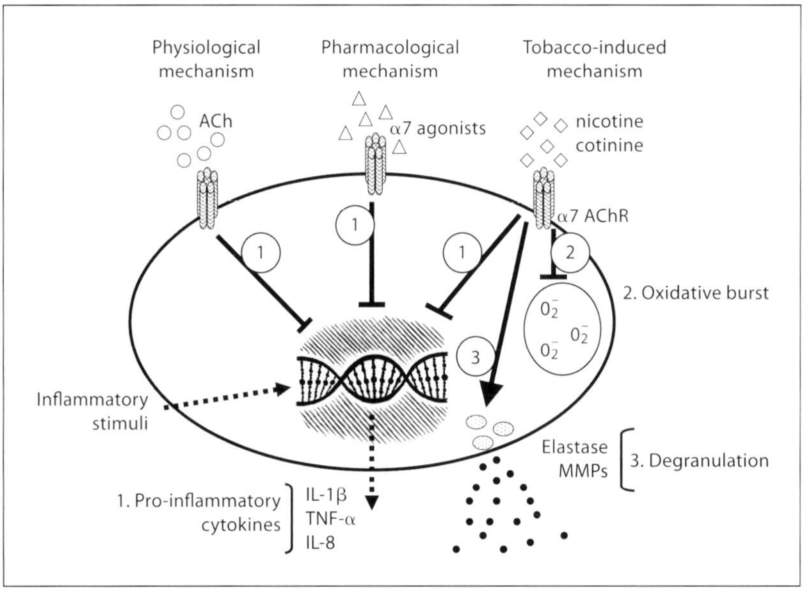

Fig. 7. Consequences of α7-nAChR engagement by various agonists in innate cells. The α7-nAChR-initiated anti-inflammatory pathway can be stimulated by various stimuli. The endogenous innate cell α7-nAChR ligand is acetylcholine that is produced by the vagus system in response to infection or other tissue insult [171, 172]. The function of acetylcholine release is to limit an overly exuberant cytokine response to infection. Many therapeutic agents are under development that attempt to take advantage of this suppressive pathway [173]. Eventually, such drugs may prove efficacious in the treatment of multiple inflammatory diseases and conditions, including septic shock, pancreatitis and inflammatory bowel diseases, as well as periodontitis [174–176]. Cigarette smoke contains large amounts of a potent exogenous α7-nAChR agonist, nicotine [163, 177–178]. The primary stable metabolite of nicotine, cotinine, is also an efficient α7-nAChR agonist [179]. The consequences of α7-nAChR engagement by acetylcholine, pharmaceutical agents or nicotine/cotinine is a suppression of the inflammatory response [174, 177, 179–180]. When triggered at the appropriate time, α7-nAChR engagement would normally be beneficial. In the case of nicotine, it has been shown that α7-nAChR engagement also results in the suppression of the oxidative burst [148, 181], a critical bacterial killing mechanism in phagocytic innate cells. Furthermore, nicotine/α7-nAChR interactions may also stimulate the rapid release of MMPs and other proteolytic enzymes [148, 152, 182] from innate cells. Therefore, chronic inappropriate nicotinic stimulation of innate α7-nAChRs in smokers may increase susceptibility to infection with periodontal pathogens by suppressing the inflammatory response and phagocytic killing required to fight infection. Additionally, increased protease release from nicotine-exposed innate cells may promote protease-protease inhibitor imbalances that contribute to the tissue remodeling apparent in periodontal diseases [183]. Figure reproduced and text adapted from Scott and Bagaitkar [146].

Anti-Neutrophil Therapeutics

As recently and extensively reviewed, a sub-antimicrobial dose of doxycycline hyclate, an antagonist of many MMP species, is the sole therapeutic designed to modulate the host response that is currently available as an adjunctive treatment for periodontitis

[184]. While such treatment is not strictly speaking neutrophil-specific, it is nevertheless pertinent to note that neutrophils represent the predominant source of destructive MMPs (MMP-8 and MMP-9) in diseased periodontal tissues [184].

Many identifiable genetic neutrophil disorders are potentially fatal; therefore, periodontal therapy is not necessarily the first priority. Because of this, and because such diseases are uncommon, studies examining the outcome of treating the underlying neutrophil defect on oral health are few in number. However, there are some encouraging data available. Neutropenias, for example, can be treated with recombinant granulocyte-colony stimulating factor or by hematopoietic stem cell transplantation [185]. A recent case report on G-CSF treatment, combined with regular maintenance therapy, in a subject with cyclic neutropenia with periodontitis proved encouraging [186]. In Papillon-Lefèvre syndrome, it has been suggested that regular maintenance therapy combined with early and aggressive antibiotic therapy can protect patients from edentulism [112].

As is appreciated by pulmonary scientists, the ideal anti-neutrophil therapeutic would take advantage of unique neutrophil surface receptors and signaling proteins, allowing manipulation of neutrophil function in a way that does not compromise antibacterial capacity [9]. Although many classes of inhibitors are under investigation, including LTB4 antagonists, CXCR2 antagonists, long-acting β2-agonists and HDAC2 activators [187], it remains to be seen whether dental research will embrace the development of neutrophil-specific therapeutic approaches to the prevention and treatment of inflammatory periodontal diseases.

Search Strategy

We used a PubMed search for 'Neutrophil OR PMN' AND 'periodont* OR gingivitis' for articles published between 2004 and 2010 that on November 5, 2010 revealed 386 hits. These data were supplemented by a selection of key articles providing contemporary hypotheses on general neutrophil physiology that were identified by 'Neutrophil OR PMN', limit 'review'.

Acknowledgement

The salary of D.A. Scott is partially funded by grant No. DE019826 from the National Institute of Dental and Craniofacial Research.

References

1 Meng H, Xu L, Li Q, Han J, Zhao Y: Determinants of host susceptibility in aggressive periodontitis. Periodontol 2000 2007;43:133–159.

2 Schenkein HA: Host responses in maintaining periodontal health and determining periodontal disease. Periodontol 2000 2006;40:77–93.

3 Ryder MI: Comparison of neutrophil functions in aggressive and chronic periodontitis. Periodontol 2000 2010;53:124–137.

4 Deas DE, Mackey SA, McDonnell HT: Systemic disease and periodontitis: manifestations of neutrophil dysfunction. Periodontol 2000 2003;32:82–104.

5 Bender JS, Thang H, Glogauer M: Novel rinse assay for the quantification of oral neutrophils and the monitoring of chronic periodontal disease. J Periodontal Res 2006;41:214–220.

6 Liu K, Cao CF, Meng HX, Gao Y: Polymorphonuclear neutrophils and their mediators in gingival tissues from generalized aggressive periodontitis. J Periodontol 2001;72:1545–1553.

7 Andersen E, Cimasoni G: A rapid and simple method for counting crevicular polymorphonuclear leucocytes. J Clin Periodontol 1993;20:651–655.

8 Borregaard N, Sorensen OE, Theilgaard-Monch K: Neutrophil granules: a library of innate immunity proteins. Trends Immunol 2007;28:340–345.

9 Cowburn AS, Condliffe AM, Farahi N, Summers C, Chilvers ER: Advances in neutrophil biology: clinical implications. Chest 2008;134:606–612.

10 Le Cabec V, Cowland JB, Calafat J, Borregaard N: Targeting of proteins to granule subsets is determined by timing and not by sorting: the specific granule protein NGAL is localized to azurophil granules when expressed in HL-60 cells. Proc Natl Acad Sci USA 1996;93:6454–6457.

11 Brumell JH, Volchuk A, Sengelov H, Borregaard N, Cieutat AM, Bainton DF, Grinstein S, Klip A: Subcellular distribution of docking/fusion proteins in neutrophils, secretory cells with multiple exocytic compartments. J Immunol 1995;155:5750–5759.

12 Borregaard N, Cowland JB: Granules of the human neutrophilic polymorphonuclear leukocyte. Blood 1997;89:3503–3521.

13 DiStasi MR, Ley K: Opening the flood-gates: how neutrophil-endothelial interactions regulate permeability. Trends Immunol 2009;30:547–556.

14 Guzik K, Skret J, Smagur J, Bzowska M, Gajkowska B, Scott DA, Potempa JS: Cigarette smoke-exposed neutrophils die unconventionally but are rapidly phagocytosed by macrophages. Cell Death Dis 2011;2:e131.

15 Borregaard N: Development of neutrophil granule diversity. Ann N Y Acad Sci 1997;832:62–68.

16 Simon SI, Cherapanov V, Nadra I, Waddell TK, Seo SM, Wang Q, Doerschuk CM, Downey GP: Signaling functions of L-selectin in neutrophils: alterations in the cytoskeleton and colocalization with CD18. J Immunol 1999;163:2891–2901.

17 Kennedy AD, DeLeo FR: Neutrophil apoptosis and the resolution of infection. Immunol Res 2009;43:25–61.

18 Hayashi F, Means TK, Luster AD: Toll-like receptors stimulate human neutrophil function. Blood 2003;102:2660–2669.

19 Quinn MT, Ammons MC, Deleo FR: The expanding role of NADPH oxidases in health and disease: no longer just agents of death and destruction. Clin Sci (Lond) 2006;111:1–20.

20 Loesche WJ, Robinson JP, Flynn M, Hudson JL, Duque RE: Reduced oxidative function in gingival crevicular neutrophils in periodontal disease. Infect Immun 1988;56:156–160.

21 Reeves EP, Lu H, Jacobs HL, Messina CG, Bolsover S, Gabella G, Potma EO, Warley A, Roes J, Segal AW: Killing activity of neutrophils is mediated through activation of proteases by K^+ flux. Nature 2002;416:291–297.

22 Pham CT: Neutrophil serine proteases fine-tune the inflammatory response. Int J Biochem Cell Biol 2008;40:1317–1333.

23 Almeida RP, Vanet A, Witko-Sarsat V, Melchior M, McCabe D, Gabay JE: Azurocidin, a natural antibiotic from human neutrophils: expression, antimicrobial activity, and secretion. Protein Expr Purif 1996;7:355–366.

24 Soehnlein O: Direct and alternative antimicrobial mechanisms of neutrophil-derived granule proteins. J Mol Med 2009;87:1157–1164.

25 Chung WO, Dommisch H, Yin L, Dale BA: Expression of defensins in gingiva and their role in periodontal health and disease. Curr Pharm Des 2007;13:3073–3083.

26 Ji S, Hyun J, Park E, Lee BL, Kim KK, Choi Y: Susceptibility of various oral bacteria to antimicrobial peptides and to phagocytosis by neutrophils. J Periodontal Res 2007;42:410–419.

27 de Haar SF, Hiemstra PS, van Steenbergen MT, Everts V, Beertsen W: Role of polymorphonuclear leukocyte-derived serine proteinases in defense against *Actinobacillus actinomycetemcomitans*. Infect Immun 2006;74:5284–5291.

28 Permpanich P, Kowolik MJ, Galli DM: Resistance of fluorescent-labelled Actinobacillus actinomycetemcomitans strains to phagocytosis and killing by human neutrophils. Cell Microbiol 2006;8:72–84.
29 Havens AM, Chiu E, Taba M, Wang J, Shiozawa Y, Jung Y, Taichman LS, D'Silva NS, Gopalakrishnan R, Wang C, Giannobile WV, Taichman RS: Stromal-derived factor-1alpha (CXCL12) levels increase in periodontal disease. J Periodontol 2008;79:845–853.
30 Page RC, Schroeder HE: Pathogenesis of inflammatory periodontal disease. A summary of current work. Lab Invest 1976;34:235–249.
31 Bainbridge B, Verma RK, Eastman C, Yehia B, Rivera M, Moffatt C, Bhattacharyya I, Lamont RJ, Kesavalu L: Role of *Porphyromonas gingivalis* phosphoserine phosphatase enzyme SerB in inflammation, immune response, and induction of alveolar bone resorption in rats. Infect Immun 2010;78:4560–4569.
32 Zaric S, Shelburne C, Darveau R, Quinn DJ, Weldon S, Taggart CC, Coulter WA: Impaired immune tolerance to *Porphyromonas gingivalis* lipopolysaccharide promotes neutrophil migration and decreased apoptosis. Infect Immun 2010;78:4151–4156.
33 Kachlany SC: *Aggregatibacter actinomycetemcomitans* leukotoxin: from threat to therapy. J Dent Res 2010;89:561–570.
34 Lally ET, Kieba IR, Sato A, Green CL, Rosenbloom J, Korostoff J, Wang JF, Shenker BJ, Ortlepp S, Robinson MK, Billings PC: RTX toxins recognize a beta2 integrin on the surface of human target cells. J Biol Chem 1997;272:30463–30469.
35 Brinkmann V, Laube B, Abu Abed U, Goosmann C, Zychlinsky A: Neutrophil extracellular traps: how to generate and visualize them. J Vis Exp 2010;36. pii:1724.
36 Logters T, Margraf S, Altrichter J, Cinatl J, Mitzner S, Windolf J, Scholz M: The clinical value of neutrophil extracellular traps. Med Microbiol Immunol 2009;198:211–219.
37 Krautgartner WD, Klappacher M, Hannig M, Obermayer A, Hartl D, Marcos V, Vitkov L: Fibrin mimics neutrophil extracellular traps in SEM. Ultrastruct Pathol 2010;34:226–231.
38 Vitkov L, Klappacher M, Hannig M, Krautgartner WD: Neutrophil fate in gingival crevicular fluid. Ultrastruct Pathol 2010;34:25–30.
39 Vitkov L, Klappacher M, Hannig M, Krautgartner WD: Extracellular neutrophil traps in periodontitis. J Periodontal Res 2009;44:664–672.
40 Buchanan JT, Simpson AJ, Aziz RK, Liu GY, Kristian SA, Kotb M, Feramisco J, Nizet V: DNase expression allows the pathogen group A Streptococcus to escape killing in neutrophil extracellular traps. Curr Biol 2006;16:396–400.
41 Potempa JG, Scott DA: Cigarette smoke-exposed neutrophils die unconventionally but are effectively phagocytosed. Cell Death Dis 2011;2:e131.
42 Watson RW, Redmond HP, Wang JH, Condron C, Bouchier-Hayes D: Neutrophils undergo apoptosis following ingestion of *Escherichia coli*. J Immunol 1996;156:3986–3992.
43 Cabrini M, Nahmod K, Geffner J: New insights into the mechanisms controlling neutrophil survival. Curr Opin Hematol 2010;17:31–35.
44 Kobayashi SD, Voyich JM, Buhl CL, Stahl RM, DeLeo FR: Global changes in gene expression by human polymorphonuclear leukocytes during receptor-mediated phagocytosis: cell fate is regulated at the level of gene expression. Proc Natl Acad Sci USA 2002;99:6901–6906.
45 Kobayashi SD, Voyich JM, Whitney AR, DeLeo FR: Spontaneous neutrophil apoptosis and regulation of cell survival by granulocyte macrophage-colony stimulating factor. J Leukoc Biol 2005;78:1408–1418.
46 Huynh ML, Fadok VA, Henson PM: Phosphatidylserine-dependent ingestion of apoptotic cells promotes TGF-beta1 secretion and the resolution of inflammation. J Clin Invest 2002;109:41–50.
47 Jewett A, Hume WR, Le H, Huynh TN, Han YW, Cheng G, Shi W: Induction of apoptotic cell death in peripheral blood mononuclear and polymorphonuclear cells by an oral bacterium, *Fusobacterium nucleatum*. Infect Immun 2000;68:1893–1898.
48 Ghosh A, Park JY, Fenno C, Kapila YL: *Porphyromonas gingivalis*, gamma interferon, and a proapoptotic fibronectin matrix form a synergistic trio that induces c-Jun N-terminal kinase 1-mediated nitric oxide generation and cell death. Infect Immun 2008;76:5514–5523.
49 Hotta K, Niwa M, Hara A, Ohno T, Wang X, Matsuno H, Kozawa O, Ito H, Kato K, Otsuka T, Matsui N, Uematsu T: The loss of susceptibility to apoptosis in exudated tissue neutrophils is associated with their nuclear factor-kappa B activation. Eur J Pharmacol 2001;433:17–27.
50 Van Zandbergen G, Gieffers J, Kothe H, Rupp J, Bollinger A, Aga E, Klinger M, Brade H, Dalhoff K, Maass M, Solbach W, Laskay T: *Chlamydia pneumoniae* multiply in neutrophil granulocytes and delay their spontaneous apoptosis. J Immunol 2004;172:1768–1776.

51 Johansson A, Claesson R, Hanstrom L, Sandstrom G, Kalfas S: Polymorphonuclear leukocyte degranulation induced by leukotoxin from *Actinobacillus actinomycetemcomitans.* J Periodontal Res 2000;35:85–92.
52 Taichman NS, Iwase M, Korchak H, Berthold P, Lally ET: Membranolytic activity of *Actinobacillus actinomycetemcomitans* leukotoxin. J Periodontal Res 1991;26(part 2):258–260.
53 Zysk G, Bejo L, Schneider-Wald BK, Nau R, Heinz H: Induction of necrosis and apoptosis of neutrophil granulocytes by *Streptococcus pneumoniae.* Clin Exp Immunol 2000;122:61–66.
54 Hermant B, Bibert S, Concord E, Dublet B, Weidenhaupt M, Vernet T, Gulino-Debrac D: Identification of proteases involved in the proteolysis of vascular endothelium cadherin during neutrophil transmigration. J Biol Chem 2003;278:14002–14012.
55 Di Gennaro A, Kenne E, Wan M, Soehnlein O, Lindbom L, Haeggstrom JZ: Leukotriene B4-induced changes in vascular permeability are mediated by neutrophil release of heparin-binding protein (HBP/CAP37/azurocidin). FASEB J 2009;23:1750–1757.
56 Wedmore CV, Williams TJ: Control of vascular permeability by polymorphonuclear leukocytes in inflammation. Nature 1981;289:646–650.
57 Petrache I, Birukova A, Ramirez SI, Garcia JG, Verin AD: The role of the microtubules in tumor necrosis factor-alpha-induced endothelial cell permeability. Am J Respir Cell Mol Biol 2003;28:574–581.
58 Strieter RM, Polverini PJ, Kunkel SL, Arenberg DA, Burdick MD, Kasper J, Dzuiba J, Van Damme J, Walz A, Marriott D, et al: The functional role of the ELR motif in CXC chemokine-mediated angiogenesis. J Biol Chem 1995;270:27348–27357.
59 Lindbom L: Regulation of vascular permeability by neutrophils in acute inflammation. Chem Immunol Allergy 2003;83:146–166.
60 Nakatani K, Takeshita S, Tsujimoto H, Kawamura Y, Sekine I: Inhibitory effect of serine protease inhibitors on neutrophil-mediated endothelial cell injury. J Leukoc Biol 2001;69:241–247.
61 Hajishengallis G: Too old to fight? Aging and its toll on innate immunity. Mol Oral Microbiol 2010;25:25–37.
62 Palmer RM, Wilson RF, Hasan AS, Scott DA: Mechanisms of action of environmental factors – tobacco smoking. J Clin Periodontol 2005;32(suppl 6):180–195.
63 Johnstone AM, Koh A, Goldberg MB, Glogauer M: A hyperactive neutrophil phenotype in patients with refractory periodontitis. J Periodontol 2007;78:1788–1794.
64 Matthews JB, Wright HJ, Roberts A, Cooper PR, Chapple IL: Hyperactivity and reactivity of peripheral blood neutrophils in chronic periodontitis. Clin Exp Immunol 2007;147:255–264.
65 Zarbock A, Ley K: Mechanisms and consequences of neutrophil interaction with the endothelium. Am J Pathol 2008;172:1–7.
66 Shapira L, Gordon B, Warbington M, Van Dyke TE: Priming effect of *Porphyromonas gingivalis* lipopolysaccharide on superoxide production by neutrophils from healthy and rapidly progressive periodontitis subjects. J Periodontol 1994;65:129–133.
67 Ashkenazi M, White RR, Dennison DK: Neutrophil modulation by *Actinobacillus actinomycetemcomitans.* II. Phagocytosis and development of respiratory burst. J Periodontal Res 1992;27:457–465.
68 Nicu EA, Van der Velden U, Everts V, Van Winkelhoff AJ, Roos D, Loos BG: Hyper-reactive PMNs in FcgammaRIIa 131 H/H genotype periodontitis patients. J Clin Periodontol 2007;34:938–945.
69 Matthews JB, Wright HJ, Roberts A, Ling-Mountford N, Cooper PR, Chapple IL: Neutrophil hyper-responsiveness in periodontitis. J Dent Res 2007;86:718–722.
70 Gustafsson A, Ito H, Asman B, Bergstrom K: Hyper-reactive mononuclear cells and neutrophils in chronic periodontitis. J Clin Periodontol 2006;33:126–129.
71 Restaino CG, Chaparro A, Valenzuela MA, Kettlun AM, Vernal R, Silva A, Puente J, Jaque MP, Leon R, Gamonal J: Stimulatory response of neutrophils from periodontitis patients with periodontal pathogens. Oral Dis 2007;13:474–481.
72 Giannobile WV: Host-response therapeutics for periodontal diseases. J Periodontol 2008;79(suppl):1592–1600.
73 Opdenakker G, Van den Steen PE, Dubois B, Nelissen I, Van Coillie E, Masure S, Proost P, Van Damme J: Gelatinase B functions as regulator and effector in leukocyte biology. J Leukoc Biol 2001;69:851–859.
74 Holzhausen M, Cortelli JR, da Silva VA, Franco GC, Cortelli SC, Vergnolle N: Protease-activated receptor-2 (PAR(2)) in human periodontitis. J Dent Res 2010;89:948–953.

75 Ujiie Y, Oida S, Gomi K, Arai T, Fukae M: Neutrophil elastase is involved in the initial destruction of human periodontal ligament. J Periodontal Res 2007;42:325–330.
76 Lamster IB, Ahlo JK: Analysis of gingival crevicular fluid as applied to the diagnosis of oral and systemic diseases. Ann N Y Acad Sci 2007;1098:216–229.
77 Hanioka T, Matsuse R, Shigemoto Y, Ojima M, Shizukuishi S: Relationship between periodontal disease status and combination of biochemical assays of gingival crevicular fluid. J Periodontal Res 2005;40:331–338.
78 Figueredo CM, Fischer RG, Gustafsson A: Aberrant neutrophil reactions in periodontitis. J Periodontol 2005;76:951–955.
79 Dommisch H, Vorderwulbecke S, Eberhard J, Steglich M, Jepsen S: SELDI-TOF-MS of gingival crevicular fluid – a methodological approach. Arch Oral Biol 2009;54:803–809.
80 Puklo M, Guentsch A, Hiemstra PS, Eick S, Potempa J: Analysis of neutrophil-derived antimicrobial peptides in gingival crevicular fluid suggests importance of cathelicidin LL-37 in the innate immune response against periodontogenic bacteria. Oral Microbiol Immunol 2008;23:328–323.
81 Dale BA, Kimball JR, Krisanaprakornkit S, Roberts F, Robinovitch M, O'Neal R, Valore EV, Ganz T, Anderson GM, Weinberg A: Localized antimicrobial peptide expression in human gingiva. J Periodontal Res 2001;36:285–294.
82 Kaner D, Bernimoulin JP, Kleber BM, Heizmann WR, Friedmann A: Gingival crevicular fluid levels of calprotectin and myeloperoxidase during therapy for generalized aggressive periodontitis. J Periodontal Res 2006;41:132–139.
83 Lorencini M, Silva JA, de la Hoz CL, Carvalho HF, Stach-Machado DR: Changes in MMPs and inflammatory cells in experimental gingivitis. Histol Histopathol 2009;24:157–166.
84 Tervahartiala T, Pirila E, Ceponis A, Maisi P, Salo T, Tuter G, Kallio P, Tornwall J, Srinivas R, Konttinen YT, Sorsa T: The in vivo expression of the collagenolytic matrix metalloproteinases (MMP-2, -8, -13, and -14) and matrilysin (MMP-7) in adult and localized juvenile periodontitis. J Dent Res 2000;79: 1969–1977.
85 Kubota T, Nomura T, Takahashi T, Hara K: Expression of mRNA for matrix metalloproteinases and tissue inhibitors of metalloproteinases in periodontitis-affected human gingival tissue. Arch Oral Biol 1996;41:253–262.
86 Xu L, Yu Z, Lee HM, Wolff MS, Golub LM, Sorsa T, Kuula H: Characteristics of collagenase-2 from gingival crevicular fluid and peri-implant sulcular fluid in periodontitis and peri-implantitis patients: pilot study. Acta Odontol Scand 2008;66:219–224.
87 Emingil G, Kuula H, Sorsa T, Atilla G: Gingival crevicular fluid matrix metalloproteinase-25 and -26 levels in periodontal disease. J Periodontol 2006;77:664–671.
88 Smith PC, Munoz VC, Collados L, Oyarzun AD: In situ detection of matrix metalloproteinase-9 (MMP-9) in gingival epithelium in human periodontal disease. J Periodontal Res 2004;39:87–92.
89 Guentsch A, Puklo M, Preshaw PM, Glockmann E, Pfister W, Potempa J, Eick S: Neutrophils in chronic and aggressive periodontitis in interaction with *Porphyromonas gingivalis* and *Aggregatibacter actinomycetemcomitans*. J Periodontal Res 2009;44:368–377.
90 Shin J, Ji S, Choi Y: Ability of oral bacteria to induce tissue-destructive molecules from human neutrophils. Oral Dis 2008;14:327–334.
91 Yamazaki T, Miyamoto M, Yamada S, Okuda K, Ishihara K: Surface protease of *Treponema denticola* hydrolyzes C3 and influences function of polymorphonuclear leukocytes. Microbes Infect 2006;8: 1758–1763.
92 Nemoto E, Kanaya S, Minamibuchi M, Shimauchi H: Cleavage of PDGF receptor on periodontal ligament cells by elastase. J Dent Res 2005;84:629–633.
93 Dias IH, Matthews JB, Chapple IL, Wright HJ, Dunston CR, Griffiths HR: Activation of the neutrophil respiratory burst by plasma from periodontitis patients is mediated by pro-inflammatory cytokines. J Clin Periodontol 2011;38:1–7.
94 Wright HJ, Chapple IL, Matthews JB, Cooper PR: *Fusobacterium nucleatum* regulation of neutrophil transcription. J Periodontal Res 2011;46:1–12.
95 Dias IH, Marshall L, Lambert PA, Chapple IL, Matthews JB, Griffiths HR: Gingipains from *Porphyromonas gingivalis* increase the chemotactic and respiratory burst-priming properties of the 77-amino-acid interleukin-8 variant. Infect Immun 2008;76:317–323.
96 Padrines M, Wolf M, Walz A, Baggiolini M: Interleukin-8 processing by neutrophil elastase, cathepsin G and proteinase-3. FEBS Lett 1994; 352:231–235.
97 Lim JK, Lu W, Hartley O, DeVico AL: N-terminal proteolytic processing by cathepsin G converts RANTES/CCL5 and related analogs into a truncated 4–68 variant. J Leukoc Biol 2006;80:1395–1404.

98 Coeshott C, Ohnemus C, Pilyavskaya A, Ross S, Wieczorek M, Kroona H, Leimer AH, Cheronis J: Converting enzyme-independent release of tumor necrosis factor alpha and IL-1beta from a stimulated human monocytic cell line in the presence of activated neutrophils or purified proteinase 3. Proc Natl Acad Sci USA 1999;96:6261–6266.
99 Bank U, Kupper B, Ansorge S: Inactivation of interleukin-6 by neutrophil proteases at sites of inflammation: protective effects of soluble IL-6 receptor chains. Adv Exp Med Biol 2000;477:431–437.
100 Wittamer V, Bondue B, Guillabert A, Vassart G, Parmentier M, Communi D: Neutrophil-mediated maturation of chemerin: a link between innate and adaptive immunity. J Immunol 2005;175:487–493. doi:175/1/487 [pii]
101 Soehnlein O, Lindbom L: Neutrophil-derived azurocidin alarms the immune system. J Leukoc Biol 2009;85:344–351. doi:jlb.0808495 [pii] 10.1189/jlb.0808495
102 Heissig B, Nishida C, Tashiro Y, Sato Y, Ishihara M, Ohki M, Gritli I, Rosenkvist J, Hattori K: Role of neutrophil-derived matrix metalloproteinase-9 in tissue regeneration. Histol Histopathol 2010;25:765–770.
103 Meyer-Hoffert U: Neutrophil-derived serine proteases modulate innate immune responses. Front Biosci 2009;14:3409–3418. doi:3462 [pii]
104 Carlsson G, Wahlin YB, Johansson A, Olsson A, Eriksson T, Claesson R, Hanstrom L, Henter JI: Periodontal disease in patients from the original Kostmann family with severe congenital neutropenia. J Periodontol 2006;77:744–751. doi:10.1902/jop.2006.050191
105 Putsep K, Carlsson G, Boman HG, Andersson M: Deficiency of antibacterial peptides in patients with morbus Kostmann: an observation study. Lancet 2002;360:1144–1149. doi:S0140-6736(02)11201-3 [pii] 10.1016/S0140-6736(02)11201-3
106 Spencer P, Fleming JE: Cyclic neutropenia: a literature review and report of case. ASDC J Dent Child 1985;52:108–113.
107 Hart TC, Atkinson JC: Mendelian forms of periodontitis. Periodontol 2000 2007;45:95–112. doi: PRD233 [pii] 10.1111/j.1600-0757.2007. 00233.x
108 Bailleul-Forestier I, Monod-Broca J, Benkerrou M, Mora F, Picard B: Generalized periodontitis associated with Chediak-Higashi syndrome. J Periodontol 2008;79:1263–1270. doi:10.1902/jop.2008.070440
109 Shibutani T, Gen K, Shibata M, Horiguchi Y, Kondo N, Iwayama Y: Long-term follow-up of periodontitis in a patient with Chediak-Higashi syndrome. A case report. J Periodontol 2000;71:1024–1028. doi: 10.1902/jop.2000.71.6.1024
110 Khocht A, Viera-Negron YE, Ameri A, Abdelsayed R: Periodontitis associated with Chediak-Higashi syndrome in a young African American male. J Int Acad Periodontol 2010;12:49–55.
111 Wani AA, Devkar N, Patole MS, Shouche YS: Description of two new cathepsin C gene mutations in patients with Papillon-Lefevre syndrome. J Periodontol 2006;77:233–237. doi:10.1902/jop.2006.050124
112 Ullbro C, Twetman S: Review Paper: Dental treatment for patients with Papillon-Lefevre syndrome (PLS). Eur Arch Paediatr Dent 2007;8(suppl 1):4–11. doi:06.32 [pii]
113 Cagli NA, Hakki SS, Dursun R, Toy H, Gokalp A, Ryu OH, Hart PS, Hart TC: Clinical, genetic, and biochemical findings in two siblings with Papillon-Lefevre Syndrome. J Periodontol 2005;76:2322–2329. doi:10.1902/jop.2005.76.12.2322
114 Fokkema SJ, Timmerman MF, van der Weijden FA, Wolffe GN, Renggli HH: A possible association of alpha1-antitrypsin deficiency with the periodontal condition in adults. J Clin Periodontol 1998;25:617–623.
115 Peterson RJ, Marsh CL: The relationship of alpha1-antitrypsin to inflammatory periodontal disease. J Periodontol 1979;50:31–35.
116 Sandholm L, Saxen L: Concentrations of serum protease inhibitors and immunoglobulins in juvenile periodontitis. J Periodontal Res 1983;18:527–533.
117 Toomarian L, Hashemi N: Periodontal manifestation of leukocyte adhesion deficiency type I. Arch Iran Med 2010;13:355–359. doi:0017 [pii] 010134/AIM.0017
118 Dababneh R, Al-Wahadneh AM, Hamadneh S, Khouri A, Bissada NF: Periodontal manifestation of leukocyte adhesion deficiency type I. J Periodontol 2008;79:764–768. doi:10.1902/jop.2008.070323
119 Parvaneh N, Mamishi S, Rezaei A, Rezaei N, Tamizifar B, Parvaneh L, Sherkat R, Ghalehbaghi B, Kashef S, Chavoshzadeh Z, Isaeian A, Ashrafi F, Aghamohammadi A: Characterization of 11 new cases of leukocyte adhesion deficiency type 1 with seven novel mutations in the ITGB2 gene. J Clin Immunol 2010;30:756–760. doi:10.1007/s10875-010-9433-2

120 Cox DP, Weathers DR: Leukocyte adhesion deficiency type 1:an important consideration in the clinical differential diagnosis of prepubertal periodontitis. A case report and review of the literature. Oral Surg Oral Med Oral Pathol Oral Radiol Endod 2008;105:86–90. doi:S1079-2104(07)00209-0 [pii] 10.1016/j.tripleo.2007.02.026

121 Giannopoulou C, Krause KH, Muller F: The NADPH oxidase NOX2 plays a role in periodontal pathologies. Semin Immunopathol 2008;30:273–278. doi:10.1007/s00281-008-0128-1

122 Dar-Odeh NS, Hayajneh WA, Abu-Hammad OA, Hammad HM, Al-Wahadneh AM, Bulos NK, Mahafzah AM, Shomaf MS, El-Maaytah MA, Bakri FG: Orofacial findings in chronic granulomatous disease: report of twelve patients and review of the literature. BMC Res Notes 2010;3:37. doi:1756-0500-3-37 [pii] 10.1186/1756-0500-3-37

123 Dimou NL, Nikolopoulos GK, Hamodrakas SJ, Bagos PG: Fcgamma receptor polymorphisms and their association with periodontal disease: a meta-analysis. J Clin Periodontol 2010;37:255–265. doi: CPE1530 [pii] 10.1111/j.1600-051X.2009.01530.x

124 Ho YP, Yang YH, Ho KY, Wu YM, Tsai CC: The association of Fcgamma receptor IIIb genetic polymorphism and susceptibility to periodontitis in Taiwanese individuals. J Clin Periodontol 2010;37:145–151. doi:CPE1507 [pii] 10.1111/j.1600-051X.2009.01507.x

125 Jones BE, Miettinen HM, Jesaitis AJ, Mills JS: Mutations of F110 and C126 of the formyl peptide receptor interfere with G-protein coupling and chemotaxis. J Periodontol 2003;74:475–484. doi: 10.1902/jop.2003.74.4.475

126 Gwinn MR, Sharma A, De Nardin E: Single nucleotide polymorphisms of the N-formyl peptide receptor in localized juvenile periodontitis. J Periodontol 1999;70:1194–1201. doi:10.1902/jop.1999.70.10.1194

127 Maney P, Walters JD: Formylpeptide receptor single nucleotide polymorphism 348T>C and its relationship to polymorphonuclear leukocyte chemotaxis in aggressive periodontitis. J Periodontol 2009;80: 1498–1505. doi:10.1902/jop.2009.090103

128 Maney P, Emecen P, Mills JS, Walters JD: Neutrophil formylpeptide receptor single nucleotide polymorphism 348T>C in aggressive periodontitis. J Periodontol 2009;80:492–498. doi:10.1902/jop.2009.080225

129 Gunji T, Onouchi Y, Nagasawa T, Katagiri S, Watanabe H, Kobayashi H, Arakawa S, Noguchi K, Hata A, Izumi Y, Ishikawa I: Functional polymorphisms of the FPR1 gene and aggressive periodontitis in Japanese. Biochem Biophys Res Commun 2007;364:7–13. doi:S0006-291X(07)02078-5 [pii] 10.1016/j.bbrc.2007.09.105

130 Demmer RT, Papapanou PN: Epidemiologic patterns of chronic and aggressive periodontitis. Periodontol 2000 2010;53:28–44. doi:PRD326 [pii] 10.1111/j.1600-0757.2009.00326.x

131 Ozturk A, Vieira AR: TLR4 as a risk factor for periodontal disease: a reappraisal. J Clin Periodontol 2009;36:279–286. doi:CPE1370 [pii] 10.1111/j.1600-051X.2009.01370.x

132 Nikolopoulos GK, Dimou NL, Hamodrakas SJ, Bagos PG: Cytokine gene polymorphisms in periodontal disease: a meta-analysis of 53 studies including 4178 cases and 4590 controls. J Clin Periodontol 2008;35:75–767. doi:CPE1298 [pii] 10.1111/j.1600-051X.2008.01298.x

133 Stein JM, Machulla HK, Smeets R, Lampert F, Reichert S: Human leukocyte antigen polymorphism in chronic and aggressive periodontitis among Caucasians: a meta-analysis. J Clin Periodontol 2008; 35:183–192. doi:CPE1189 [pii] 10.1111/j.1600-051X.2007.01189.x

134 Yoshie H, Kobayashi T, Tai H, Galicia JC: The role of genetic polymorphisms in periodontitis. Periodontol 2000 2007;43:102–132. doi:PRD164 [pii] 10.1111/j.1600-0757.2006.00164.x

135 Saftig P, Beertsen W, Eskelinen EL: LAMP-2:a control step for phagosome and autophagosome maturation. Autophagy 2008;4:510–512. doi:5724 [pii]

136 Beertsen W, Willenborg M, Everts V, Zirogianni A, Podschun R, Schroder B, Eskelinen EL, Saftig P: Impaired phagosomal maturation in neutrophils leads to periodontitis in lysosomal-associated membrane protein-2 knockout mice. J Immunol 2008; 180:475–482. doi:180/1/475 [pii]

137 Shiohara M, Gombart AF, Sekiguchi Y, Hidaka E, Ito S, Yamazaki T, Koeffler HP, Komiyama A: Phenotypic and functional alterations of peripheral blood monocytes in neutrophil-specific granule deficiency. J Leukoc Biol 2004;75:190–197.doi:10.1189/jlb.0203063 jlb.0203063 [pii]

138 Nibali L, Parkar M, Brett P, Knight J, Tonetti MS, Griffiths GS: NADPH oxidase (CYBA) and FcgammaR polymorphisms as risk factors for aggressive periodontitis: a case-control association study. J Clin Periodontol 2006;33:529–539. doi: CPE952 [pii] 10.1111/j.1600-051X.2006.00952.x

139 Dhanrajani PJ: Papillon-Lefevre syndrome: clinical presentation and a brief review. Oral Surg Oral Med Oral Pathol Oral Radiol Endod 2009;108:e1–e7. doi:S1079-2104(09)00165-6 [pii] 10.1016/j.tripleo.2009.03.016

140 Fransson C, Berglundh T, Lindhe J: The effect of age on the development of gingivitis. Clinical, microbiological and histological findings. J Clin Periodontol 1996;23:379–385.

141 Siegel B, Weihe E, Bette M, Nusing RM, Flores-de-Jacoby L, Mengel R: The effect of age on prostaglandin-synthesizing enzymes in the development of gingivitis. J Periodontal Res 2007;42:259–266. doi:JRE942 [pii] 10.1111/j.1600-0765.2006.00942.x
142 Butcher S, Chahel H, Lord JM: Review article: ageing and the neutrophil: no appetite for killing? Immunology 2000;100:411–416. doi:imm079 [pii]
143 Gomez CR, Nomellini V, Faunce DE, Kovacs EJ: Innate immunity and aging. Exp Gerontol 2008;43:718–728. doi:S0531-5565(08)00163-0 [pii] 10.1016/j.exger.2008.05.016
144 Guntsch A, Erler M, Preshaw PM, Sigusch BW, Klinger G, Glockmann E: Effect of smoking on crevicular polymorphonuclear neutrophil function in periodontally healthy subjects. J Periodontal Res 2006;41:184–188. doi:JRE852 [pii] 10.1111/j.1600-0765.2005.00852.x
145 Morozumi T, Kubota T, Sugita N, Itagaki M, Yoshie H: Alterations of gene expression in human neutrophils induced by smoking cessation. J Clin Periodontol 2004;31:1110–1116. doi:CPE612 [pii] 10.1111/j.1600-051X.2004.00612.x
146 Scott DA, Bagaitkar J: Smoking and innate immune (dys)function. In Bernhard D (ed): Cigarette Smoke Toxicity. Linking Individual Chemicals to Human Diseases. Wiley-VCH Verlag GmbH & Co. KGaA, Weinheim, Germany, 2011.
147 Terashima T, Klut ME, English D, Hards J, Hogg JC, van Eeden SF: Cigarette smoking causes sequestration of polymorphonuclear leukocytes released from the bone marrow in lung microvessels. Am J Respir Cell Mol Biol 1999;20:171–177.
148 Xu M, Scott JE, Liu KZ, Bishop HR, Renaud DE, Palmer RM, Soussi-Gounni A, Scott DA: The influence of nicotine on granulocytic differentiation – inhibition of the oxidative burst and bacterial killing and increased matrix metalloproteinase-9 release. BMC Cell Biol 2008;9:19.
149 Pitzer JE, Del Zoppo GJ, Schmid-Schonbein GW: Neutrophil activation in smokers. Biorheology 1996;33:45–58.
150 Selby C, Drost E, Brown D, Howie S, MacNee W: Inhibition of neutrophil adherence and movement by acute cigarette smoke exposure. Experimental Lung Research 1992;18:813–827.
151 Hind CR, Joyce H, Tennent GA, Pepys MB, Pride NB: Plasma leucocyte elastase concentrations in smokers. Journal of Clinical Pathology 1991;44:232–235.
152 Seow WK, Thong YH, Nelson RD, MacFarlane GD, Herzberg MC: Nicotine-induced release of elastase and eicosanoids by human neutrophils. Inflammation 1994;18:119–127.
153 Nakamura T, Ebihara I, Shimada N, Koide H: Effect of cigarette smoking on plasma metalloproteinase-9 concentration. Clinica Chimica Acta 1998;276:173–177. doi:S0009-8981(98)00104-1 [pii]
154 Liu KZ, Hynes A, Man A, Alsagheer A, Singer DL, Scott DA: Increased local matrix metalloproteinase-8 expression in the periodontal connective tissues of smokers with periodontal disease. Biochim Biophys Acta 2006;1762:775–780.
155 Guzik K, Skret J, Smagur J, Bzowska M, Gajkowska B, Scott DA, Potempa JS: Cigarette smoke-exposed neutrophils die unconventionally but are rapidly phagocytosed by macrophages. Cell Death and Disease, in press, 2011.
156 Zappacosta B, Persichilli S, Minucci A, Stasio ED, Carlino P, Pagliari G, Giardina B, Sole PD: Effect of aqueous cigarette smoke extract on the chemiluminescence kinetics of polymorphonuclear leukocytes and on their glycolytic and phagocytic activity. Luminescence 2001;16:315–319.
157 Benhammou K, Lee M, Strook M, Sullivan B, Logel J, Raschen K, Gotti C, Leonard S: [(3)H]Nicotine binding in peripheral blood cells of smokers is correlated with the number of cigarettes smoked per day. Neuropharmacology 2000;39:2818–2829. doi:S0028390800001532 [pii]
158 Lee HY, Kehrli ME Jr, Brogden KA, Gallup JM, Ackermann MR: Influence of beta(2)-integrin adhesion molecule expression and pulmonary infection with Pasteurella haemolytica on cytokine gene expression in cattle. Infection & Immunity 2000;68:4274–4281.
159 Fredriksson M, Bergstrom K, Asman B: IL-8 and TNF-alpha from peripheral neutrophils and acute-phase proteins in periodontitis. Journal of Clinical Periodontology 2002;29:123–128. doi:cpe290206 [pii]
160 Zeilhofer HU, Schorr W: Role of interleukin-8 in neutrophil signaling. Current Opinion in Hematology 2000;7:178–182.
161 Scott DA, Stapleton JA, Wilson RF, Sutherland G, Palmer RM, Coward PY, Gustavsson G: Dramatic decline in circulating intercellular adhesion molecule-1 concentration on quitting tobacco smoking. Blood Cells Mol Dis 2000;26:255–258.
162 Scott DA, Wilson RF, Coward PY, Odell EW, Poston RN, Matthews JP, Palmer RM: The acute influence of tobacco smoking on adhesion molecule expression on monocytes and neutrophils and on circulating adhesion molecule levels in vivo. Addiction Biology 2000;5:195–205.

163 Rehani K, Scott DA, Renaud D, Hamza H, Williams LR, Wang H, Martin M: Cotinine-induced convergence of the cholinergic and PI3 kinase-dependent anti-inflammatory pathways in innate immune cells. Biochim Biophys Acta 2008;1783:375–382.
164 Ulloa L, Tracey KJ: The 'cytokine profile': a code for sepsis. Trends Mol Med 2005;11:56–63.
165 Huston JM, Tracey KJ: The pulse of inflammation: heart rate variability, the cholinergic anti-inflammatory pathway and implications for therapy. J Intern Med 2011;269:45–53. doi:10.1111/j.1365-2796.2010.02321.x
166 Su X, Matthay MA, Malik AB: Requisite role of the cholinergic alpha7 nicotinic acetylcholine receptor pathway in suppressing Gram-negative sepsis-induced acute lung inflammatory injury. Journal of Immunology 2010;184:401–410. doi:jimmunol.0901808 [pii] 10.4049/jimmunol 0901808
167 Huston JM, Rosas-Ballina M, Xue X, Dowling O, Ochani K, Ochani M, Yeboah MM, Chatterjee PK, Tracey KJ, Metz CN: Cholinergic neural signals to the spleen down-regulate leukocyte trafficking via CD11b. Journal of Immunology 2009;183:552–559. doi:183/1/552 [pii] 10.4049/jimmunol.0802684
168 Sadis C, Teske G, Stokman G, Kubjak C, Claessen N, Moore F, Loi P, Diallo B, Barvais L, Goldman M, Florquin S, Le Moine A: Nicotine protects kidney from renal ischemia/reperfusion injury through the cholinergic anti-inflammatory pathway. PLoS One 2007;2:e469. doi:10.1371/journal.pone.0000469
169 Giebelen IA, van Westerloo DJ, LaRosa GJ, de Vos AF, van der Poll T: Stimulation of alpha 7 cholinergic receptors inhibits lipopolysaccharide-induced neutrophil recruitment by a tumor necrosis factor alpha-independent mechanism. Shock 2007;27:443–447. doi:10.1097/01.shk.0000245016.78493.bb00024382-200704000-00017 [pii]
170 van Westerloo DJ, Giebelen IA, Florquin S, Daalhuisen J, Bruno MJ, de Vos AF, Tracey KJ, van der Poll T: The cholinergic anti-inflammatory pathway regulates the host response during septic peritonitis. Journal of Infectious Diseases 2005;191: 2138–2148. doi:JID33743 [pii] 10.1086/ 430323
171 Borovikova LV, Ivanova S, Zhang M, Yang H, Botchkina GI, Watkins LR, Wang H, Abumrad N, Eaton JW, Tracey KJ: Vagus nerve stimulation attenuates the systemic inflammatory response to endotoxin. Nature 2000;405:458–462.
172 Huston JM, Ochani M, Rosas-Ballina M, Liao H, Ochani K, Pavlov VA, Gallowitsch-Puerta M, Ashok M, Czura CJ, Foxwell B, Tracey KJ, Ulloa L: Splenectomy inactivates the cholinergic antiinflammatory pathway during lethal endotoxemia and polymicrobial sepsis. J Exp Med 2006;203:1623–1628.
173 Rosas-Ballina M, Goldstein RS, Gallowitsch-Puerta M, Yang L, Valdes-Ferrer SI, Patel NB, Chavan S, Al-Abed Y, Yang H, Tracey KJ: The selective alpha7 agonist GTS-21 attenuates cytokine production in human whole blood and human monocytes activated by ligands for TLR2, TLR3, TLR4, TLR9, and RAGE. Mol Med 2009;15:195–202.
174 van Westerloo DJ, Giebelen IA, Florquin S, Bruno MJ, Larosa GJ, Ulloa L, Tracey KJ, van der Poll T: The vagus nerve and nicotinic receptors modulate experimental pancreatitis severity in mice. Gastroenterology 2006;130:1822–1830.
175 Scott DA, Martin M: Exploitation of the nicotinic anti-inflammatory pathway for the treatment of epithelial inflammatory diseases. World J Gastroenterol 2006;12:7451–7459.
176 Pavlov VA, Ochani M, Yang LH, Gallowitsch-Puerta M, Ochani K, Lin X, Levi J, Parrish WR, Rosas-Ballina M, Czura CJ, Larosa GJ, Miller EJ, Tracey KJ, Al-Abed Y: Selective alpha7-nicotinic acetylcholine receptor agonist GTS-21 improves survival in murine endotoxemia and severe sepsis. Crit Care Med 2007;35:1139–1144.
177 Matsunaga K, Klein TW, Friedman H, Yamamoto Y: Involvement of nicotinic acetylcholine receptors in suppression of antimicrobial activity and cytokine responses of alveolar macrophages to Legionella pneumophila infection by nicotine. J Immunol 2001;167:6518–6524.
178 de Heens GL, Kikkert R, Aarden LA, van der Velden U, Loos BG: Effects of smoking on the ex vivo cytokine production in periodontitis. J Periodontal Res 2009;44:28–34.
179 Rehani K, Wang H, Garcia CA, Kinane DF, Martin M: Toll-like receptor-mediated production of IL-1Ra is negatively regulated by GSK3 via the MAPK ERK1/2. J Immunol 2009;182:547–553.
180 Wang H, Yu M, Ochani M, Amella CA, Tanovic M, Susarla S, Li JH, Wang H, Yang H, Ulloa L, Al-Abed Y, Czura CJ, Tracey KJ: Nicotinic acetylcholine receptor alpha7 subunit is an essential regulator of inflammation. Nature 2003;421:384–388.
181 Pabst MJ, Pabst KM, Collier JA, Coleman TC, Lemons-Prince ML, Godat MS, Waring MB, Babu JP: Inhibition of neutrophil and monocyte defensive functions by nicotine. J Periodontol 1995;66:1047–1055.
182 Murphy EA, Danna-Lopes D, Sarfati I, Rao SK, Cohen JR: Nicotine-stimulated elastase activity release by neutrophils in patients with abdominal aortic aneurysms. Annals of Vascular Surgery 1998; 12:41–45.

183 Cheng SL, Wang HC, Yu CJ, Tsao PN, Carmeliet P, Cheng SJ, Yang PC: Prevention of elastase-induced emphysema in placenta growth factor knock-out mice. Respir Res 2009;10:115. doi:1465–9921–10–115 [pii] 10.1186/1465–9921–10–115.
184 Preshaw PM: Host response modulation in periodontics. Periodontol 2000 2008;48:92–110. doi: PRD252 [pii] 10.1111/j.1600–0757.2008. 00252.x
185 Klein C: Congenital neutropenia. Hematology Am Soc Hematol Educ Program 2009:344–350. doi: 2009/1/344 [pii] 10.1182/asheducation-2009. 1.344
186 Matarasso S, Daniele V, Iorio Siciliano V, Mignogna MD, Andreuccetti G, Cafiero C: The effect of recombinant granulocyte colony-stimulating factor on oral and periodontal manifestations in a patient with cyclic neutropenia: a case report. Int J Dent 2009, 654239 (2009) doi:10.1155/2009/654239
187 Burgos RA, Hidalgo MA, Figueroa CD, Conejeros I, Hancke JL: New potential targets to modulate neutrophil function in inflammation. Mini Rev Med Chem 2009;9:153–168.

David A. Scott, PhD
Center for Oral Health and Systemic Disease
501 South Preston
Louisville, KY 40202–1701 (USA)
Tel. +1 502 852 8905, E-Mail dascot07@louisville.edu

Kinane DF, Mombelli A (eds): Periodontal Disease. Front Oral Biol. Basel, Karger, 2012, vol 15, pp 84–98

Antimicrobial Peptides in Periodontal Innate Defense

Sven-Ulrik Gorr

Department of Diagnostic and Biological Sciences, University of Minnesota School of Dentistry, Minneapolis, Minn., USA

Abstract

The development of oral biofilms and the host response to biofilm bacteria and their toxins are important factors in the development of periodontal disease. An early component of the host response is the secretion of antimicrobial proteins and peptides (AMPs) by salivary glands, oral epithelial cells and neutrophils. Over 45 AMPs have been identified in the oral cavity. All are found in saliva and several are also present in gingival crevicular fluid. Of these, 13 are upregulated in periodontal disease while 11 are downregulated. However, the concentrations of most AMPs found in oral fluids are below the effective in vitro concentrations, suggesting that local concentrations must be higher for effect or that additional biological functions are important in the oral cavity. Thus, in addition to direct antibacterial activity (e.g. bactericidal activity, bacterial agglutination), AMPs may affect the course of periodontal disease by inactivating bacterial or host proteases (e.g. secretory leukoprotease inhibitor) or bind bacterial toxins, including lipopolysaccharides (e.g. LL-37). Several AMPs (e.g. defensins) also act as immune system alarmins, i.e. endogenous mediators that recruit and activate antigen-presenting cells to enhance innate and adaptive immune responses. The differential regulation of AMP expression in periodontal disease suggests that AMP panels, including up- and downregulated proteins, can be used in oral fluid diagnosis of periodontal disease and to monitor treatment outcome.

Periodontal disease is an inflammatory disease initiated by the formation of mixed biofilms on the teeth and gingival tissues. Biofilm bacteria and their toxins perturb gingival epithelial cells leading to a sequence of inflammatory and immune processes that may result in the destruction of gingival tissues, attachment loss and periodontal pockets. In susceptible patients, alveolar bone loss and tooth loss may follow as a result of periodontal disease. At least 50% of adults exhibit some level of periodontal disease [1, 2].

The oral cavity contains over 700 resident species of bacteria, of which 150–200 bacterial species are typically found in individuals. About 400 species of bacteria have been found in periodontal pockets, but a small subset of these are found in individual subjects [3]. In fact, only 8 bacterial species have consistently been associated with periodontal disease, including *A. actinomycetemcomitans, P. gingivalis, T. forsythia, T. denticola, F. nucleatum, E. nodatum, P. intermedia and P. nigrescens* [4–6]. *P. gingivalis, T. forsythia and T. denticola* belong to the red complex described by Socransky et al. [5], while *A. actinomycetemcomitans, P. gingivalis* and *T. forsythia* are considered consensus pathogens [4]. However, these bacteria are found in both healthy and diseased sites, although in much higher numbers (up to 10^5-fold) in the latter sites [7]. Thus, it has been proposed that, rather than the bacterial infection per se, the magnitude of the inflammatory response is a determinant of the development of destructive periodontitis [8]. It appears that the development of oral biofilms and the host response to biofilm bacteria and their toxins are important factors in the progression to periodontal disease. The host response consists of a cascade of events by the innate and acquired immune systems. An early component of this cascade is the secretion of antimicrobial proteins and peptides (AMPs) by salivary glands, oral epithelial cells and neutrophils.

Antimicrobial Peptides and Oral Bacteria

A complex mixture of over 45 AMPs are found in oral fluids. In addition, recent salivary proteome data suggest that additional AMPs will be found in oral fluids [9]. Thus, over 1,100 proteins have been identified in human saliva and a large number of these are proteins of unknown function. The known AMPs belong to 6 functional families [10]: (1) cationic peptides, (2) bacterial agglutination and adhesion, (3) metal ion chelators, (4) peroxidases, (5) protease inhibitors, and (6) AMPs with activity against bacterial cell walls. It is thought that this variety of proteins and functions allows the oral innate immune system to defend against the wide variety of microbes that enter the mouth. Moreover, the variety of functions appears to be a successful strategy against bacterial AMP resistance. Interestingly, the absence of a single AMP may lead to increased periodontal disease. Thus, morbus Kostmann disease is characterized by severe neutropenia and periodontal disease. In particular, the lack of LL-37 has been associated with periodontal disease since treatments to restore neutrophil levels did not clear the inflammation. However, in a single patient, a bone marrow transplant restored neutrophil levels, LL-37 levels and gingival health [11]. Similarly, Papillon-Lefèvre syndrome is characterized by increased susceptibility to early-onset periodontitis and is caused by mutations in the cathepsin C gene. The lack of active cathepsin prevents the proteolytic processing of cathelicidin to the active AMP LL-37 [12, 13].

Despite the large number and variety of oral AMPs, it appears that the innate immune system is unable to cope with the normal oral flora, including oral pathogens associated with periodontal disease. Thus, in the absence of regular oral care, teeth and gingival tissues quickly acquire biofilms that, if left untreated, cause gingivitis [14]. Interestingly, the associated inflammation may be linked to modern diets since subjects adhering to a stone age diet, in the absence of oral care, develop supragingival plaque but this does not lead to an increase in the severity of gingival inflammation, bleeding on probing or pocket depths [15]. Thus, it is possible that (refined sugars in) modern diets contribute to vigorous growth of biofilms and hence overwhelm natural innate defenses.

In contrast to the ability of oral biofilms to develop in the presence of oral AMPs, tooth extraction, oral surgery or other acute injuries to the oral mucosa rarely lead to infection or damaging inflammation. It is possible that the neutrophils recruited to the site of injury provide the necessary local dose of AMPs to prevent bacterial colonization.

Expression of Antimicrobial Proteins and Peptides in the Oral Cavity

The AMPs found in saliva and gingival crevicular fluid (GCF) have recently been reviewed [10, 16]. Although the data are not complete, current evidence suggests that only a handful of AMPs, including adrenomedullin, elafin, human neutrophil peptide (HNP)-1 and statherin, are present in saliva and GCF in concentrations higher than the minimal inhibitory concentration (MIC) for oral bacteria [16]. Given the localized nature of periodontal disease, it is important to note that the saliva concentration of AMPs may be significantly diluted, and that this may not be directly relevant to the concentration that would be found in the periodontal pocket. Thus, it cannot be excluded that locally higher concentrations are reached at the site of secretion from oral epithelial cells or neutrophils. Especially neutrophils are recruited to sites of injury and secrete a rich mixture of AMPs (table 1). While this mixture of AMPs is effective against planktonic bacteria, it may nevertheless be inadequate against bacterial biofilms. Thus, when encountering a mature biofilm, the neutrophils may be unable to engulf biofilm bacteria, leading to a 'frustrated phagocytosis' causing the release of enzymes and products of the oxidative burst into the periodontal pocket where they can cause tissue destruction.

Over 45 AMPs have been identified in the oral cavity [10, 16]. All are found in saliva and several are also present in GCF. Of these, 13 are upregulated in periodontal disease while 11 are downregulated (table 2). As an example, defensins and LL-37, key components of the mucosal antimicrobial defense, are induced by bacterial stimulation of gingival epithelial cells while only LL-37 expression is induced in peripheral blood neutrophils [17]. In some cases where AMPs were measured in saliva or GCF, downregulation may be caused by proteolytic degradation of the AMP by bacterial proteases. It is not clear if the regulation of AMP expression is a functional response to the bacterial insult or if it is part of a more general stimulation of target cells. In fact, even the stimulated levels of several AMPs are below the MIC for oral pathogens [16]. Thus, other biological activities of the AMPs may play a functional role in gingival inflammation.

Table 1. Oral AMPs (saliva or GCF) stored in neutrophil granules

Gene	Protein
AZU1	azurocidin
B2M	β_2-microglobulin
BPI	bactericidal/permeability-increasing protein
CAMP	cathelicidin (LL-37)
DEFA1	HNP-1 (neutrophil defensin)
DEFA1, DEFA3	HNP-2
DEFA3	HNP-3
LTF	lactoferrin
LYZ	lysozyme
MPO	myeloperoxidase
PGLYRP1	peptidoglycan recognition protein 1
S100A7	psoriasin, protein S100-A7
SLPI	secretory leukoprotease inhibitor protein

Neutrophil data are from Lehrer and Ganz [85] and Lominadze et al. [86].

Biological Roles of Antibacterial Peptides and Proteins

AMPs exhibit multiple biological activities that could play a role in the innate immune defense against oral bacteria and their toxins. In addition to direct antibacterial activity (e.g. bactericidal activity, bacterial agglutination), AMPs may affect the course of periodontal disease by inactivating bacterial or host proteases (secretory leukoprotease inhibitor; SLPI) or bind bacterial toxins, including lipopolysaccharide (LPS) (LL-37). Several AMPs affect the inflammatory response and wound healing in mucosal tissues and can act as modulators of the immune system [18–20]. As examples of these activities, the biological activities of LL-37, secretory leukoprotease inhibitor and parotid secretory protein (PSP) are reviewed.

Cathelicidin (LL-37)
LL-37 is a multifunctional 37-amino-acid peptide derived from human cathelicidin by proteolytic processing following secretion [21]. The peptide is a random coil in aqueous solution, but assumes an α-helical conformation when exposed to lipid membranes. The peptide is expressed in epithelial cells and neutrophils, and is found in saliva at 0.14–3 μg/ml [22, 23]. LL-37 has been detected in GCF by immunoblotting and ELISA (0.5 μg/ml) and the levels are increased in chronic periodontitis [24, 25]. Within individual patients, LL-37 levels are higher in GCF from inflamed sites compared with healthy sites [26], and LL-37 deficiency has been directly implicated in the development of periodontitis [11].

The peptide is active against gram-negative and gram-positive bacteria, including *A. actinomycetemcomitans* and *P. gingivalis*. Activity (LD_{99}) against three strains of *A.*

Table 2. Antimicrobial proteins that are differentially regulated in periodontal disease

Protein	Gene	Change in periodontitis	Sample	Reference
Upregulated in periodontal disease				
Adrenomedullin	*ADM*	up 2-fold	GCF	87
β_2-Microglobulin	*B2M*	up 3- to 10-fold	GCF	88
β-Defensin-1, hBD1	*DEFB1*	upregulated chronic	mRNA gingiva	89
β-Defensin-4A β-Defensin-2 hBD-2	*DEFB4A*	upregulated aggressive	mRNA gingiva	89
Calgranulin A Protein S100-A8	*S100A8*	up 2-3-fold	GCF	90
Cathelicidin (LL-37)	*CAMP*	upregulated aggressive and chronic periodontitis	GCF	24
Cystatin C	*CST3*	upregulated	whole saliva	91
Hemoglobin β-Globin α-Globin	*HBB* *HBA1 HBA2*	increased due to bleeding		
HNP 1–3 Neutrophil defensins 1–3	*DEFA1*	up 15-fold aggressive periodontitis up 60-fold chronic periodontitis	GCF	24
Lysozyme C	*LYZ*	up aggressive periodontitis	GCF	92
Short palate, lung and nasal epithelium clone 2 PSP	*SPLUNC2*	up 3.3-fold	whole saliva	67
Transferrin Serotransferrin	*TF*	upregulated	GCF	93
Vasoactive intestinal peptide	*VIP*	upregulated	GCF	83
Downregulated in periodontal disease				
β-Defensin-1, hBD1	*DEFB1*	downregulated aggressive	mRNA gingiva	89
β-Defensin 103 β-Defensin-3, hBD3	*DEFB103A*	downregulated		94
Calcitonin gene-related peptide 1	*CALCA*	downregulated	GCF	95
Cystatin S	*CST4*	downregulated	whole saliva	91

Table 2. Continued

Protein	Gene	Change in periodontitis	Sample	Reference
Cystatin SA	*CST2*	downregulated	whole saliva	96
Cystatin SN	*CST1*	downregulated	whole saliva	97
Fibronectin	*FN1*	down 2-fold with less intact fibronectin in periodontitis	GCF	98 99
Lactoferrin	*LTF*	down 1.7-fold aggressive (highly variable)	whole saliva	67
Lactotransferrin				100
Mucin 7. MG2	*MUC7*	down 3-fold	saliva	100
Neuropeptide Y	*NPY*	downregulated in periodontitis	GCF	101
Secretory leukoprotease inhibitor protein	*SLPI*	downregulated in periodontitis	GCF	58

When known, the fold-change, whether the change occurs in aggressive or chronic periodontitis, and the sample site are presented.

actinomycetemcomitans is about 10 μg/ml and is inhibited by 100 mM sodium chloride [27]. In a study of LL-37 activity under physiological conditions, it was noted that the inhibition by salt was mitigated by the presence of carbonate in physiological buffers. Thus, target bacteria exhibit modifications to their cell wall in the presence of carbonate that result in increased susceptibility to LL-37 under physiological conditions [28].

The MICs against different strains of *A. actinomycetemcomitans* are 30–60 μg/ml and >125 μg/ml against *P. gingivalis* [29]. The latter result may be due to proteolytic degradation of LL-37 by gingipains secreted by *P. gingivalis*. Saliva components protect the peptide against degradation by gingipains released by the periodontal pathogen *P. gingivalis*. While the antibacterial activity of LL-37 is lowered by saliva, the net effect in vivo appears to be the protection of antibacterial activity in the presence of *P. gingivalis* proteases [30].

The concentrations of LL-37 in saliva and GCF are lower than the MIC for oral bacteria, suggesting that local concentrations must be higher for effect or that other biological functions are important in the oral cavity. The high concentrations of LL-37 needed for antibacterial activity and the inactivation by serum proteins have led to the suggestion that the primary function of this and other AMPs is as an immune system alarmin, i.e. an endogenous mediator that recruits and activates antigen-presenting cells to enhance innate and adaptive immune responses [19]. Indeed, LL-37 acts as a chemoattractant for monocytes, T cells and neutrophils. The effect is inhibited by pertussis toxin suggesting that it involves a G_i-protein-coupled receptor, which was shown to be formyl peptide receptor-like 1 [31].

LL-37 has also been implicated in carcinogenesis in breast, ovarian and lung cancers, but suppresses tumorigenesis in gastric cancer. Thus, the effects of LL-37 on tumorigenesis are tissue specific [32]. In lung cancer cells, LL-37 activates MAP kinase pathways and increases their proliferation and anchorage-independent growth [33].

Mucus plays an important role in protecting mucosal surfaces from microbial attack. In colonic epithelium, LL-37 causes an increase in mucus thickness and expression of mucin genes [34]. The upregulation of mucin expression is mediated by phosphorylation of MAP kinase, which is upregulated by LL-37 in epithelial HT-29 cells [35]. An effect of LL-37 on mucin expression in oral epithelial cells has not been reported, but mucin inhibits the antibacterial activity of LL-37 [36].

LL-37 directly binds LPS [37] and inhibits the inflammatory response in human gingival fibroblasts stimulated with heat-killed *P. gingivalis*, bacterial LPS or *P. gingivalis* fimbriae. In each case, the expression of IL6, IL8 and CXCL10 were inhibited [38].

Taken together, LL-37 is a multifunctional peptide that could affect the development of periodontal disease from bacterial colonization, bacterial toxicity and host response.

Defensins

The defensin family are prominent AMPs in oral epithelial cells and neutrophils [39]. Human β-defensins (hBD) are primarily expressed in epithelial cells, while α-defensins (HNP) are predominantly expressed in neutrophils. Levels of HNP1–3 in GCF are upregulated 60-fold in chronic periodontitis and 15-fold in aggressive periodontitis [24]. The level reached in chronic periodontitis (70 μg/ml) is above the MIC for *S. mutans* [40], but the peptides are not active against *A. actinomycetemcomitans* or *P. gingivalis* [41, 42].

The biological activities of defensins overlap with those of LL-37. Defensins are stored in secretory granules at a high concentration, and may kill phagocytosed bacteria upon fusion of secretory granules with phagocytic vacuoles [39]. As for LL-37, the antibacterial effect of defensins is salt sensitive, and it has been suggested that other biological activities constitute the primary function of secreted defensins. Indeed, defensins also act as alarmins that both chemoattract and activate antigen-presenting cells [19]. As an example, hBD-2, but not hBD-1, acts as a chemoattractant for mast cells at a concentration of 3 μg/ml [43], i.e. similar to the concentration reached in periodontitis and significantly lower than the MIC for oral bacteria [16].

Defensins inhibit LPS-stimulated inflammatory responses in host cells. As an example, human β-defensins and HNPs (except HNP-2) inhibit the expression of IL-1ß, IL-8 and ICAM-1 in gingival fibroblast stimulated with *P. intermedia* LPS [44].

Defensin expression in gingival epithelial cells from periodontal patients is differentially regulated by periodontal pathogens and commensals [45]. Differences were also seen when hBD expression was induced by different capsule types of *P. gingivalis* or different serotypes of *A. actinomycetemcomitans* [45]. Periodontitis has a genetic component of about 50% [46], and a recent study has found a significant association of periodontal disease with a SNP of the 3′ untranslated region of the *DEFB1* gene (hBD-1) [47].

In addition to the potential role of defensins in oral infection and inflammation, an unusual role for the canine β-defensin gene *CBD103* was recently reported in domestic dogs, where CBD 103 interacts with melanocortin 1 receptor in determining black coat color [48]. Interestingly, in mice another coat color gene locus – agouti – has been associated with *Psp* [49, 50], a proposed host defense gene (see below).

Secretory Leukoprotease Inhibitor

SLPI is an 11.7-kD non-glycosylated serine-protease inhibitor that was originally isolated from parotid saliva [51]. The protein is expressed in neutrophils, epithelial cells and salivary glands, and inhibits serine proteases, including neutrophil elastase, cathepsin G and trypsin, which are involved in inflammatory tissue destruction [52]. SLPI is also expressed in oral tumor tissues [53]. In addition to its protease inhibitor activity, salivary SLPI inhibits HIV-1 infection of human monocytes [54] and the protein is fungicidal against *A. fumigatus* and *C. albicans* [55].

Salivary concentrations of SLPI are 0.1–10 μg/ml [52, 56, 57]. The amount of SLPI in GCF increased 3-fold from about 80 to 300 ng/ml following periodontal treatment [58].

SLPI is associated with wound repair and its expression is increased following wounding. Accordingly, *Slpi*-null mice are defective in cutaneous wound healing and exhibit increased elastase activity and inflammation. The latter is associated with increased activation of TGFβ in the knockout animals [59]. SLPI is secreted by macrophages and increased amounts are secreted when the cells are stimulated with apoptotic cells [60]. Together these findings suggest that SLPI plays a role in the resolution of inflammation. This is consistent with the increased expression of SLPI following periodontal treatment [58].

Parotid Secretory Protein

Parotid secretory protein (PSP) (short palate, lung and nasal epithelium clone-2, SPLUNC2) was originally detected in mouse and rat salivary glands [61, 62]. Human PSP was identified in acinar and ductal epithelial cells of the parotid gland [63, 64], and the secreted protein has been detected in whole, parotid and submandibular/sublingual saliva [9, 63, 65]. PSP is expressed in human gingival epithelial cells and its expression is increased by heat-inactivated *P. gingivalis* [66] and upregulated in saliva from generalized aggressive periodontitis patients compared to healthy controls [67].

PSP belongs to a family of oral and airway proteins that are related to palate, lung and nasal epithelium clone (PLUNC) [68]. Based on a predicted similarity to bactericidal/permeability-increasing protein and LPS-binding protein, PSP and the other members of the PLUNC protein family have been proposed to play a role in host defense [69]. Indeed, mouse PSP binds bacteria [70] and peptides derived from the human PSP sequence cause bacterial agglutination [71]. The related proteins human PLUNC and bovine BSP30 inhibit growth of *Mycoplasma pneumoniae* [72] and *Pseudomonas aeruginosa* and *Streptococcus pneumoniae* [73], respectively.

Since PLUNC proteins are related to LPS-binding proteins, the PLUNC family members human PLUNC and bovine BSP30 have also been tested for LPS binding, but with varying results. Ghafouri et al. [74] reported that PLUNC binds LPS, whereas no significant binding was detected for BSP30 [73]. Recent experiments with human salivary PSP demonstrated that this protein binds immobilized LPS and this binding was specifically inhibited by a PSP peptide [75]. Together, these findings support the proposed host defense function of PSP.

As is the case for other AMPs, the PLUNC proteins also exhibit multiple functions. Thus, PLUNC and the horse protein latherin have surfactant properties that may aid in the wetting of mucosal and dermal surfaces [76, 77]. Based on sequence similarities, similar functions can be predicted for PSP and the other members of the PLUNC protein family. A specific role for PSP or other PLUNC proteins in the development of periodontitis has not yet been reported.

Antimicrobial Peptides and Proteins as Biomarkers for Periodontal Disease

Saliva and GCF hold promise as diagnostic tools for both oral and systemic diseases [78, 79]. The gingival changes associated with periodontitis, including changes in the expression of cytokines and AMPs, and products of tissue and bone destruction are potential targets for oral fluid diagnostics and recent reviews have identified salivary biomarkers for periodontal disease from these protein categories [80, 81]. The potential markers include proteins involved in inflammation (e.g. C-reactive protein, IL-1β, TNFα), collagen breakdown (MMP-8, MMP-9), bone remodeling (alkaline phosphatase, osteocalcin, osteonectin) and host defense proteins (e.g. lysozyme, lactoferrin, histatin).

Although the saliva levels of individual markers can be statistically distinguished between healthy samples and periodontitis, the large variation in individual sample values make a prospective assignment of any given saliva marker more difficult. A possible solution to this problem is the use of diagnostic panels that combine several markers [81]. Based on the known regulation of AMPs in periodontal disease, we propose that this class of proteins, which are typically expressed in response to oral bacteria or bacterial toxins, are an attractive target as markers for periodontal disease. We have previously identified 45 AMPs in saliva and GCF [10, 16]. Further analysis of this data set identified 13 AMPs that are upregulated in periodontal disease and 11 AMPs that are downregulated (table 2). Based on these data, we propose that panels of AMPs may be suitable for the diagnosis of periodontal disease in saliva samples. The use of ratios between up- and downregulated proteins from individual patients may overcome individual variation and potentiate the observed differences to allow prospective assignment of any given sample as 'healthy' or 'diseased'. Since saliva diagnoses the overall oral condition, a positive test could then be followed by analysis of GCF from suspected sites or diagnosis by the periodontal practitioner. In support of this approach, the AMPs PSP, prolactin-inducible protein, vasoactive intestinal peptide and myeloperoxidase are decreased after therapy, while SLPI and calgranulins are increased [58, 82–84]. Thus, treatment outcomes may be monitored by changes in AMP levels pre- and post-treatment.

Summary

A large number of AMPs have been identified in the oral cavity and several are regulated by and express antimicrobial activity to periodontal pathogens. However, the concentrations of these proteins found in oral fluids are often below the effective in vitro concentrations. Thus, it is thought that AMPs act intracellularly or at the site of secretion from neutrophils and epithelial cells or act through alternate biological functions, including acting as alarmins, anti-inflammatory proteins and as signaling molecules for resolution of inflammation and wound healing. It is proposed that the differential regulation of AMPs in periodontal disease make them attractive biomarkers for the disease in saliva and GCF.

Acknowledgements

The author's work on Parotid Secretory Protein and antimicrobial peptides was supported by USPHS grant numbers 2R01 DE012205 and 1R01 DE017989 from the National Institute for Dental and Craniofacial Research at the National Institutes of Health. Additional support from the University of Louisville and University of Minnesota Schools of Dentistry is gratefully acknowledged.

References

1 Morris AJ, Steele J, White DA: The oral cleanliness and periodontal health of UK adults in 1998. Br Dent J 2001;191:186–192.

2 König J, Holtfreter B, Kocher T: Periodontal health in Europe: future trends based on treatment needs and the provision of periodontal services – position paper 1. Eur J Dent Educ 2010;14:4–24.

3 Paster BJ, Olsen I, Aas JA, Dewhirst FE: The breadth of bacterial diversity in the human periodontal pocket and other oral sites. Periodontol 2000 2006; 42:80–87.

4 Periodontics WWoC: Consensus report. Periodontal diseases: pathogenesis and microbial factors. Ann Periodontol 1996;1:926–932.

5 Socransky SS, Haffajee AD, Cugini MA, Smith C, Kent RL Jr: Microbial complexes in subgingival plaque. J Clin Periodontol 1998;25:134–144.

6 Teles RP, Haffajee AD, Socransky SS: Microbiological goals of periodontal therapy. Periodontol 2000 2006; 42:180–218.

7 Lovegrove JM: Dental plaque revisited: Bacteria associated with periodontal disease. J N Z Soc Periodontol 2004:7–21.

8 Van Dyke TE: Control of inflammation and periodontitis. Periodontology 2000 2007;45:158–166.

9 Denny P, Hagen FK, Hardt M, Liao L, Yan W, Arellanno M, Bassilian S, Bedi GS, Boontheung P, Cociorva D, Delahunty CM, Denny T, Dunsmore J, Faull KF, Gilligan J, Gonzalez-Begne M, Halgand F, Hall SC, Han X, Henson B, Hewel J, Hu S, Jeffrey S, Jiang J, Loo JA, Ogorzalek Loo RR, Malamud D, Melvin JE, Miroshnychenko O, Navazesh M, Niles R, Park SK, Prakobphol A, Ramachandran P, Richert M, Robinson S, Sondej M, Souda P, Sullivan MA, Takashima J, Than S, Wang J, Whitelegge JP, Witkowska HE, Wolinsky L, Xie Y, Xu T, Yu W, Ytterberg J, Wong DT, Yates JR, Fisher SJ: The proteomes of human parotid and submandibular/sublingual gland salivas collected as the ductal secretions. J Proteome Res 2008;7:1994–2006.

10 Gorr S-U: Antimicrobial peptides of the oral cavity. Periodontol 2000 2009;51:152–180.

11 Putsep K, Carlsson G, Boman HG, Andersson M: Deficiency of antibacterial peptides in patients with morbus Kostmann: an observation study. Lancet 2002;360:1144–1149.

12 de Haar SF, Hiemstra PS, van Steenbergen MTJM, Everts V, Beertsen W: Role of polymorphonuclear leukocyte-derived serine proteinases in defense against actinobacillus actinomycetemcomitans. Infect Immun 2006;74:5284–5291.

13 Hart TC, Hart PS, Michalec MD, Zhang Y, Firatli E, Van Dyke TE, Stabholz A, Zlotogorski A, Shapira L, Soskolne WA: Haim-Munk syndrome and Papillon-Lefevre syndrome are allelic mutations in cathepsin C. J Med Genet 2000;37:88–94.

14 Loe H, Theilade E, Jensen SB: Experimental gingivitis in man. J Periodontol 1965;36:177–187.

15 Baumgartner S, Imfeld T, Schicht O, Rath C, Persson RE, Persson GR: The impact of the stone age diet on gingival conditions in the absence of oral hygiene. J Periodontol 2009;80:759–768.

16 Gorr SU, Abdolhosseini M: Antimicrobial peptides and periodontal disease. J Clin Periodontol 2011;38: 126–141.

17 Hosokawa I, Hosokawa Y, Komatsuzawa H, Goncalves RB, Karimbux N, Napimoga MH, Seki M, Ouhara K, Sugai M, Taubman MA, Kawai T: Innate immune peptide ll-37 displays distinct expression pattern from beta-defensins in inflamed gingival tissue. Clin Exp Immunol 2006;146:218–225.

18 Diamond G, Beckloff N, Weinberg A, Kisich KO: The roles of antimicrobial peptides in innate host defense. Curr Pharm Des 2009;15:2377–2392.

19 Yang D, Oppenheim JJ: Alarmins and antimicrobial immunity. Med Mycol 2009;47:S146–S153.

20 Soehnlein O: Direct and alternative antimicrobial mechanisms of neutrophil-derived granule proteins. J Mol Med 2009;87:1157–1164.

21 Kai-Larsen Y, Agerberth B: The role of the multifunctional peptide ll-37 in host defense. Front Biosci 2008;13:3760–3767.

22 Bachrach G, Chaushu G, Zigmond M, Yefenof E, Stabholz A, Shapira J, Merrick J, Chaushu S: Salivary ll-37 secretion in individuals with down syndrome is normal. J Dent Res 2006;85:933–936.

23 Tao R, Jurevic RJ, Coulton KK, Tsutsui MT, Roberts MC, Kimball JR, Wells N, Berndt J, Dale BA: Salivary antimicrobial peptide expression and dental caries experience in children. Antimicrob Agents Chemother 2005;49:3883–3888.

24 Puklo M, Guentsch A, Hiemstra PS, Eick S, Potempa J: Analysis of neutrophil-derived antimicrobial peptides in gingival crevicular fluid suggests importance of cathelicidin ll-37 in the innate immune response against periodontogenic bacteria. Oral Microbiol Immunol 2008;23:328–335.

25 Turkoglu O, Emingil G, Kutukculer N, Atilla G: Gingival crevicular fluid levels of cathelicidin ll-37 and interleukin-18 in patients with chronic periodontitis. J Periodontol 2009;80:969–976.

26 Dommisch H, Vorderwülbecke S, Eberhard J, Steglich M, Jepsen S: SELDI–TOF-MS of gingival crevicular fluid – a methodological approach. Arch Oral Biol 2009;54:803–809.
27 Tanaka D, Miyasaki KT, Lehrer RI: Sensitivity of *Actinobacillus actinomycetemcomitans* and *Capnocytophaga* spp. to the bactericidal action of LL-37: a cathelicidin found in human leukocytes and epithelium. Oral Microbiol Immunol 2000;15:226–231.
28 Dorschner RA, Lopez-Garcia B, Peschel A, Kraus D, Morikawa K, Nizet V, Gallo RL: The mammalian ionic environment dictates microbial susceptibility to antimicrobial defense peptides. FASEB J 2006;20:35–42.
29 Ji S, Hyun J, Park E, Lee BL, Kim KK, Choi Y: Susceptibility of various oral bacteria to antimicrobial peptides and to phagocytosis by neutrophils. J Periodontal Res 2007;42:410–419.
30 Gutner M, Chaushu S, Balter D, Bachrach G: Saliva enables the antimicrobial activity of LL-37 in the presence of proteases of porphyromonas gingivalis. Infect Immun 2009;77:5558–5563.
31 Yang D, Chen Q, Schmidt AP, Anderson GM, Wang JM, Wooters J, Oppenheim JJ, Chertov O: LL-37, the neutrophil granule- and epithelial cell-derived cathelicidin, utilizes formyl peptide receptor-like 1 (FPRL1) as a receptor to chemoattract human peripheral blood neutrophils, monocytes, and T cells. J Exp Med 2000;192:1069–1074.
32 Wu WKK, Wang G, Coffelt SB, Betancourt AM, Lee CW, Fan D, Wu K, Yu J, Sung JJY, Cho CH: Emerging roles of the host defense peptide LL-37 in human cancer and its potential therapeutic applications. Int J Cancer 2010;127:1741–1747.
33 von Haussen J, Koczulla R, Shaykhiev R, Herr C, Pinkenburg O, Reimer D, Wiewrodt R, Biesterfeld S, Aigner A, Czubayko F, Bals R: The host defence peptide LL-37/hCAP-18 is a growth factor for lung cancer cells. Lung Cancer 2008;59:12–23.
34 Tai EKK, Wu WKK, Wong HPS, Lam EKY, Yu L, Cho CH: A new role for cathelicidin in ulcerative colitis in mice. Exp Biol Med 2007;232:799–808.
35 Tai EKK, Wong HPS, Lam EKY, Wu WKK, Yu L, Koo MWL, Cho CH: Cathelicidin stimulates colonic mucus synthesis by up-regulating MUC1 and MUC2 expression through a mitogen-activated protein kinase pathway. J Cell Biochem 2008; 104:251–258.
36 Bucki R, Namiot DB, Namiot Z, Savage PB, Janmey PA: Salivary mucins inhibit antibacterial activity of the cathelicidin-derived LL-37 peptide but not the cationic steroid CSA-13. J Antimicrob Chemother 2008;62:329–225.
37 Larrick JW, Hirata M, Balint RF, Lee J, Zhong J, Wright SC: Human cap18: a novel antimicrobial lipopolysaccharide-binding protein. Infect Immun 1995;63:1291–1297.
38 Inomata M, Into T, Murakami Y: Suppressive effect of the antimicrobial peptide LL-37 on expression of IL-6, IL-8 and CXCL10 induced by *Porphyromonas gingivalis* cells and extracts in human gingival fibroblasts. Eur J Oral Sci 2010;118:574–581.
39 Ganz T: Defensins: antimicrobial peptides of innate immunity. Nat Rev Immunol 2003;3:710–720.
40 Lundy FT, Nelson J, Lockhart D, Greer B, Harriott P, Marley JJ: Antimicrobial activity of truncated [alpha]-defensin (human neutrophil peptide (HNP)-1) analogues without disulphide bridges. Mol Immunol 2008;45:190–193.
41 Raj PA, Antonyraj KJ, Karunakaran T: Large-scale synthesis and functional elements for the antimicrobial activity of defensins. Biochem J 2000;347(part 3):633–641.
42 Miyasaki KT, Bodeau AL, Ganz T, Selsted ME, Lehrer RI: In vitro sensitivity of oral, gram-negative, facultative bacteria to the bactericidal activity of human neutrophil defensins. Infect Immun 1990;58:3934–3940.
43 Niyonsaba F, Iwabuchi K, Matsuda H, Ogawa H, Nagaoka I: Epithelial cell-derived human β-defensin-2 acts as a chemotaxin for mast cells through a pertussis toxin-sensitive and phospholipase c-dependent pathway. Int Immunol 2002;14:421–426.
44 Lee SH, Jun HK, Lee HR, Chung CP, Choi BK: Antibacterial and lipopolysaccharide (LPS)-neutralising activity of human cationic antimicrobial peptides against periodontopathogens. Int J Antimicrob Agents 2010;35:138–145.
45 Vankeerberghen A, Nuytten H, Dierickx K, Quirynen M, Cassiman JJ, Cuppens H: Differential induction of human beta-defensin expression by periodontal commensals and pathogens in periodontal pocket epithelial cells. J Periodontol 2005;76:1293–1303.
46 Michalowicz BS, Diehl SR, Gunsolley JC, Sparks BS, Brooks CN, Koertge TE, Califano JV, Burmeister JA, Schenkein HA: Evidence of a substantial genetic basis for risk of adult periodontitis. J Periodontol 2000;71:1699–1707.
47 Schaefer AS, Richter GM, Nothnagel M, Laine ML, Ruhling A, Schafer C, Cordes N, Noack B, Folwaczny M, Glas J, Dorfer C, Dommisch H, Groessner-Schreiber B, Jepsen S, Loos BG, Schreiber S: A 3′ UTR transition within DEFB1 is associated with chronic and aggressive periodontitis. Genes Immun 2009;11:45–54.

48 Candille SI, Kaelin CB, Cattanach BM, Yu B, Thompson DA, Nix MA, Kerns JA, Schmutz SM, Millhauser GL, Barsh GS: A β-defensin mutation causes black coat color in domestic dogs. Science 2007;318:1418–1423.
49 Owerbach D, Hjorth JP: Inheritance of a parotid secretory protein in mice and its use in determining salivary amylase quantitative variants. Genetics 1980;95:129–141.
50 Barsh GS, Epstein CJ: Physical and genetic-characterization of a 75-kilobase deletion associated with al, a recessive lethal allele at the mouse agouti locus. Genetics 1989;121:811–818.
51 Thompson RC, Ohlsson K: Isolation, properties, and complete amino acid sequence of human secretory leukocyte protease inhibitor, a potent inhibitor of leukocyte elastase. Proc Natl Acad Sci U S A 1986; 83:6692–6696.
52 Doumas S, Kolokotronis A, Stefanopoulos P: Anti-inflammatory and antimicrobial roles of secretory leukocyte protease inhibitor. Infect Immun 2005;73: 1271–1274.
53 Westin U, Nystrom M, Ljungcrantz I, Eriksson B, Ohlsson K: The presence of elafin, SLPI, IL1-RA and STNFalpha RI in head and neck squamous cell carcinomas and their relation to the degree of tumour differentiation. Mediators Inflamm 2002; 11:7–12.
54 McNeely TB, Dealy M, Dripps DJ, Orenstein JM, Eisenberg SP, Wahl SM: Secretory leukocyte protease inhibitor: a human saliva protein exhibiting anti-human immunodeficiency virus 1 activity in vitro. J Clin Invest 1995;96:456–464.
55 Tomee JF, Hiemstra PS, Heinzel-Wieland R, Kauffman HF: Antileukoprotease: an endogenous protein in the innate mucosal defense against fungi. J Infect Dis 1997;176:740–747.
56 Shugars DC, Watkins CA, Cowen HJ: Salivary concentration of secretory leukocyte protease inhibitor, an antimicrobial protein, is decreased with advanced age. Gerontology 2001;47:246–253.
57 Lin AL, Johnson DA, Stephan KT, Yeh C-K: Salivary secretory leukocyte protease inhibitor increases in HIV infection. J Oral Pathol Med 2004;33:410–416.
58 Nakamura-Minami M, Furuichi Y, Ishikawa K, Mitsuzono-Tofuku Y, Izumi Y: Changes of alpha1-protease inhibitor and secretory leukocyte protease inhibitor levels in gingival crevicular fluid before and after non-surgical periodontal treatment. Oral Dis 2003;9:249–254.
59 Ashcroft GS, Lei K, Jin W, Longenecker G, Kulkarni AB, Greenwell-Wild T, Hale-Donze H, McGrady G, Song X-Y, Wahl SM: Secretory leukocyte protease inhibitor mediates non-redundant functions necessary for normal wound healing. Nat Med 2000;6: 1147–1153.
60 Odaka C, Mizuochi T, Yang J, Ding A: Murine macrophages produce secretory leukocyte protease inhibitor during clearance of apoptotic cells: implications for resolution of the inflammatory response. J Immunol 2003;171:1507–1514.
61 Madsen HO, Hjorth JP: Molecular cloning of mouse PSP mRNA. Nucleic Acids Res 1985;13:1–13.
62 Mirels L, Ball WD: Neonatal rat submandibular-gland protein SMG-A and parotid secretory protein are alternatively regulated members of a salivary protein multigene family. J Biol Chem 1992;267: 2679–2687.
63 Geetha C, Venkatesh SG, Fasciotto Dunn BH, Gorr S-U: Expression and anti-bacterial activity of human parotid secretory protein (PSP). Biochem Soc Trans 2003;31:815–818.
64 Geetha C, Venkatesh SG, Bingle L, Bingle CD, Gorr SU: Design and validation of anti-inflammatory peptides from human parotid secretory protein. J Dent Res 2005;84:149–153.
65 Bingle L, Barnes FA, Lunn H, Musa M, Webster S, Douglas CW, Cross SS, High AS, Bingle CD: Characterisation and expression of SPLUNC2, the human orthologue of rodent parotid secretory protein. Histochem Cell Biol 2009;132:339–349.
66 Shiba H, Venkatesh SG, Gorr SU, Barbieri G, Kurihara H, Kinane DF: Parotid secretory protein is expressed and inducible in human gingival keratinocytes. J Periodontal Res 2005;40:153–157.
67 Wu Y, Shu R, Luo LJ, Ge LH, Xie YF: Initial comparison of proteomic profiles of whole unstimulated saliva obtained from generalized aggressive periodontitis patients and healthy control subjects. J Periodontal Res 2009;44:636–644.
68 Bingle CD, Craven CJ: PLUNC: a novel family of candidate host defence proteins expressed in the upper airways and nasopharynx. Hum Mol Genet 2002;11:937–943.
69 Bingle CD, Gorr S-U: Host defense in oral and airway epithelia: chromosome 20 contributes a new protein family. Int J Biochem Cell Biol 2004;36:2144–2152.
70 Robinson CP, Bounous DI, Alford CE, Nguyen KH, Nanni JM, Peck AB, Humphreys-Beher MG: PSP expression in murine lacrimal glands and function as a bacteria binding protein in exocrine secretions. Am J Physiol 1997;272:G863–G871.

71 Gorr S-U, Sotsky JB, Shelar AP, Demuth DR: Design of bacteria-agglutinating peptides derived from parotid secretory protein, a member of the bactericidal/permeability increasing-like protein family. Peptides 2008;29:2118–2127.
72 Chu HW, Thaikoottathil J, Rino JG, Zhang G, Wu Q, Moss T, Refaeli Y, Bowler R, Wenzel SE, Chen Z, Zdunek J, Breed R, Young R, Allaire E, Martin RJ: Function and regulation of SPLUNC1 protein in mycoplasma infection and allergic inflammation. J Immunol 2007;179:3995–4002.
73 Haigh B, Hood K, Broadhurst M, Medele S, Callaghan M, Smolenski G, Dines M, Wheeler T: The bovine salivary proteins BSP30a and BSP30b are independently expressed BPI-like proteins with anti-Pseudomonas activity. Mol Immunol 2007;45:1944–1951.
74 Ghafouri B, Kihlstrom E, Tagesson C, Lindahl M: PLUNC in human nasal lavage fluid: multiple isoforms that bind to lipopolysaccharide. Biochim Biophys Acta 2004;1699:57–63.
75 Abdolhosseini M, Sotsky JB, Shelar AP, Joyce PB, Gorr SU: Human parotid secretory protein is a lipopolysaccharide-binding protein: identification of an anti-inflammatory peptide domain. Mol Cell Biochem 2011;DOI 10.1007/s11010-011-0991-2.
76 Gakhar L, Bartlett JA, Penterman J, Mizrachi D, Singh PK, Mallampalli RK, Ramaswamy S, McCray PB Jr: PLUNC is a novel airway surfactant protein with anti-biofilm activity. PLoS One 2010;5:e9098.
77 McDonald RE, Fleming RI, Beeley JG, Bovell DL, Lu JR, Zhao X, Cooper A, Kennedy MW: Latherin: a surfactant protein of horse sweat and saliva. PLoS ONE 2009;4:e5726.
78 Lee YH, Wong DT: Saliva: an emerging biofluid for early detection of diseases. Am J Dent 2009;22:241–248.
79 Malamud D: Saliva as a diagnostic fluid. Dental Clin N Am 2011;55:159–178.
80 Giannobile WV, Beikler T, Kinney JS, Ramseier CA, Morelli T, Wong DT: Saliva as a diagnostic tool for periodontal disease: current state and future directions. Periodontology 2000 2009;50:52–64.
81 Miller CS, Foley JD, Bailey AL, Campell CL, Humphries RL, Christodoulides N, Floriano PN, Simmons G, Bhagwandin B, Jacobson JW, Redding SW, Ebersole JL, McDevitt JT: Current developments in salivary diagnostics. Biomark Med 2010;4:171–189.
82 Haigh BJ, Stewart KW, Whelan JR, Barnett MP, Smolenski GA, Wheeler TT: Alterations in the salivary proteome associated with periodontitis. J Clin Periodontol 2010;37:241–247.
83 Linden GJ, Mullally BH, Burden DJ, Lamey PJ, Shaw C, Ardill J, Lundy FT: Changes in vasoactive intestinal peptide in gingival crevicular fluid in response to periodontal treatment. J Clin Periodontol 2002;29:484–489.
84 Kaner D, Bernimoulin J, Kleber B, Heizmann W, Friedmann A: Gingival crevicular fluid levels of calprotectin and myeloperoxidase during therapy for generalized aggressive periodontitis. J Periodontal Res 2006;41:132–139.
85 Lehrer RI, Ganz T: Antimicrobial polypeptides of human neutrophils. Blood 1990;76:2169–2181.
86 Lominadze G, Powell DW, Luerman GC, Link AJ, Ward RA, McLeish KR: Proteomic analysis of human neutrophil granules. Mol Cell Proteomics 2005;4:1503–1521.
87 Lundy FT, O'Hare MMT, McKibben BM, Fulton CR, Briggs JE, Linden GJ: Radioimmunoassay quantification of adrenomedullin in human gingival crevicular fluid. Arch Oral Biol 2006;51:334–338.
88 Mogi M, Otogoto J, Ota N, Inagaki H, Minami M, Kojima K: Interleukin 1[beta], interleukin 6, [beta]2-microglobulin, and transforming growth factor-[alpha] in gingival crevicular fluid from human periodontal disease. Arch Oral Biol 1999;44:535–539.
89 Vardar-Sengul S, Demirci T, Sen BH, Erkizan V, Kurulgan E, Baylas H: Human β-defensin-1 and -2 expression in the gingiva of patients with specific periodontal diseases. J Periodontal Res 2007;42:429–437.
90 Lundy FT, Chalk R, Lamey P-J, Shaw C, Linden GJ: Quantitative analysis of MRP-8 in gingival crevicular fluid in periodontal health and disease using microbore HPLC. J Clin Periodontol 2001;28:1172–1177.
91 Henskens YM, van den Keijbus PA, Veerman EC, Van der Weijden GA, Timmerman MF, Snoek CM, Van der Velden U, Nieuw Amerongen AV: Protein composition of whole and parotid saliva in healthy and periodontitis subjects: determination of cystatins, albumin, amylase and IgA. J Periodontal Res 1996;31:57–65.
92 Friedman SA, Mandel ID, Herrera MS: Lysozyme and lactoferrin quantitation in the crevicular fluid. J Periodontol 1983;54:347–350.
93 Adonogianaki E, Mooney J, Kinane DF: The ability of gingival crevicular fluid acute phase proteins to distinguish healthy, gingivitis and periodontitis sites. J Clin Periodontol 1992;19:98–102.
94 Bissell J, Joly S, Johnson GK, Organ CC, Dawson D, McCray PB Jr, Guthmiller JM: Expression of β-defensins in gingival health and in periodontal disease. J Oral Pathol Med 2004;33:278–285.

95 Lundy FT, Shaw C, McKinnell J, Lamey PJ, Linden GJ: Calcitonin gene-related peptide in gingival crevicular fluid in periodontal health and disease. J Clin Periodontol 1999;26:212–216.
96 Ito T, Komiya-Ito A, Arataki T, Furuya Y, Yajima Y, Yamada S, Okuda K, Kato T: Relationship between antimicrobial protein levels in whole saliva and periodontitis. J Periodontol 2008;79:316–322.
97 Baron AC, Gansky SA, Ryder MI, Featherstone DB: Cysteine protease inhibitory activity and levels of salivary cystatins in whole saliva of periodontally diseased patients. J Periodontal Res 1999;34:437–444.
98 Lopatin DE, Caffesse ER, Bye FL, Caffesse RG: Concentrations of fibronectin in the sera and crevicular fluid in various stages of periodontal disease. J Clin Periodontol 1989;16:359–364.
99 Wang PL, Azuma Y, Shinohara M, Ohura K: Effect of actinobacillus actinomycetemcomitans protease on the proliferation of gingival epithelial cells. Oral Dis 2001;7:233–237.
100 Groenink J, Walgreen-Weterings E, Nazmi K, Bolscher JG, Veerman EC, van Winkelhoff AJ, Nieuw Amerongen AV: Salivary lactoferrin and low-Mr mucin MG2 in *Actinobacillus actinomycetemcomitans*-associated periodontitis. J Clin Periodontol 1999;26: 269–275.
101 Lundy FT, El Karim IA, Linden GJ: Neuropeptide Y (NPY) and NPY Y1 receptor in periodontal health and disease. Arch Oral Biol 2009;54:258–262.

Sven-Ulrik Gorr, PhD
Department of Diagnostic and Biological Sciences
University of Minnesota School of Dentistry
Minneapolis, MN 55455 (USA)
Tel. +1 612 626 5728, E-Mail sugorr@umn.edu

Kinane DF, Mombelli A (eds): Periodontal Disease. Front Oral Biol. Basel, Karger, 2012, vol 15, pp 99–116

Modern Approaches to Non-Surgical Biofilm Management

Danae Anastasia Apatzidou

Department of Preventive Dentistry, Periodontology and Biology of Implants, Dental School, Aristotle University of Thessaloniki, Thessaloniki, Greece

Abstract

The subgingival dental plaque is a microbial biofilm consisting of highly variable bacterial microcolonies embedded within a self-produced matrix of extracellular polymeric substance. In contrast to microorganisms growing in a planktonic state, the inhabitants of a biofilm are effectively protected within this dense structure from host defense mechanisms and from therapeutic agents, including antimicrobials. The mechanical removal of the microbial biofilm and the establishment of meticulous plaque control measures comprise the key elements for the success of non-surgical periodontal treatment. Ultrasonic devices are effective in disrupting the biofilm, and carefully remove soft and hard deposits from a root surface with minimal trauma to the tooth structure. Controversies and modern trends in non-surgical periodontal therapy – such as quadrant-wise treatment modalities versus full-mouth approaches, hand- versus power-driven instrumentation, and the time frame of non-surgical periodontal therapy – are discussed here in depth in order to provide an insight into modern approaches to non-surgical biofilm management. Clinical, microbiological and immunological findings following different treatment protocols, in addition to cost-effective benefits of these clinical modalities, are discussed.

General Properties of the Biofilm Related to Therapeutic Issues

The biofilm consists of highly variable bacterial microcolonies embedded within a self-produced matrix of extracellular polymeric substance, and lines the tooth surface on the one hand and the epithelial cells of the periodontium on the other hand. A zone of microorganisms that are loosely attached or might be in a planktonic state is lodged in the middle of this double-layer structure and connects the two parts of the biofilm [1]. The tooth- and epithelial-cell-associated sections have distinct properties in their physiological state, composition and proportions of various bacterial species, and in their response to treatment procedures. The first line of host defense involves neutrophil migration out of a blood vessel in the periodontal tissues into the lamina propria, and through there into the junctional epithelium, and eventually into the gingival crevice/periodontal pocket, where a wall of neutrophils is formed against the surface layer of the biofilm, trying to neutralize bacteria in the plaque biofilm [Scott and Krauss, pp. 56–83]. It has been shown that bacteria within biofilms are in a dynamic state and interact with each other by direct cell-to-cell contact, produce signaling molecules, and exchange genetic information and nutrients with other residents of the environment [Zijnge et al., pp. 1–16; Diaz, pp. 17–40].

In contrast to microorganisms growing in a planktonic state, the inhabitants of a biofilm are effectively protected within this dense film from the host response and from any other therapeutical agent, including antimicrobials. Detachment of cells from biofilms to colonize novel sites is an important element of the continuing struggle for survival of the microbial species. In order to achieve this, the pathogen should be able to attach to the host tissues, multiply, compete with other subgingival species and finally escape from the host defense mechanisms. There are a number of difficulties a microorganism may encounter in colonizing a periodontal pocket. The constant outflow of the gingival crevicular fluid and the presence of saliva washing out the tissues tend to prevent the attachment of pathogens. In addition, the constituents of saliva and gingival crevicular fluid have many innate and adaptive antimicrobial properties. The anatomy of the local tissues, desquamation of the epithelial cells and mechanical masticatory forces are innate processes that limit aggregation of dental plaque on the gingivae and provide defensive mechanisms against the pathogens. These parameters are comprehensively discussed in other chapters in this volume [Benakanakere and Kinane, pp. 41–55; Scott and Krauss, pp. 56–83; Gorr, pp. 84–98]. Bacteria in dental plaque have evolved to overcome these obstacles and translocate from one site to another in the same or another individual. The microbes which re-colonize a subgingival pocket could be either residual microorganisms following incomplete instrumentation or the extension of a growing and maturing supragingival plaque.

A number of studies question the role that supragingival plaque control has on the levels of pathogens within deep periodontal pockets [2–5], implying that the periodontal pocket is a separate ecological niche independent from the supragingival environment, and that the nutrients for the subgingival microorganisms are supplied by other members of the microflora or by the gingival crevicular fluid. It has been suggested that the bacteria detected in the pockets following instrumentation, rather than the supragingival plaque control measures, play a more important role in the re-establishment of the subgingival microbial flora [6]. On the other hand, evidence shows that optimal supragingival plaque control can inhibit the microbial re-population of subgingival pockets that have been previously periodontally treated [7–10], and seems to affect the microbial composition at sites that have not received any additional subgingival instrumentation [11–15]. It is of interest to note that in the presence of supragingival plaque deposits, it takes approximately 4–8 weeks for the subgingival microflora to shift back to pretreatment values [7, 16]. In conclusion, there is a general agreement in the literature that at periodontally diseased sites with an established subgingival ecosystem, meticulous supragingival plaque control remains the cornerstone of successful periodontal treatment.

Biological Rationale for Mechanical Instrumentation

The initial objective of scaling and root planing was to remove calculus and bacterial deposits, as well as cementum contaminated by bacteria and their products [17–19], based on the concept that bacterial endotoxins are not loosely adhering to the surface of the exposed root cementum, but instead penetrate into the cementum and leach back to periodontal tissues [20–21]. Nonetheless, this theory has not been validated by more recent studies [22–24] and, in contrast to the traditional belief, periodontal health can be achieved without the removal of the 'infected' root cementum by scaling and root planing [25]. Animal studies also support this finding [26–28]. Although there is no biological rationale for the intentional excessive removal of cementum in order to produce a smooth root surface devoid of bacterial endotoxins, root planing should not be ruled out since calculus cannot be eliminated without the removal of cementum. Hughes et al. [23] suggested that the primary therapeutic advantage of root planing is that it facilitates the removal of plaque and calculus from root irregularities such as resorption lacunae, which seem to be a prominent feature of the affected root surfaces. The finding that a large amount of endotoxins can be removed by comparatively gentle procedures implies that effective root surface debridement can be achieved by relatively simple and non-traumatic methods other than using traditional hand instrumentation [22]. The authors suggested that an ultrasonic device might be used for non-surgical debridement, not as a scaling instrument, but by simply passing it lightly over the root surfaces to remove bacteria and their products.

On this basis, the use of instruments that disrupt the biofilm and carefully remove soft and hard deposits from a periodontal site, while causing minimal mechanical trauma to the structure of the root, is considered highly advantageous, and ultrasonic devices are effective to this end [29, 30]. Numerous studies have demonstrated that manual and sonic/ultrasonic instrumentation are equally efficacious in treating periodontal disease, in that both approaches demonstrate similar efficiency in subgingival calculus removal and plaque control, and also exhibit a similar healing response [31–38]. In addition, it should be noted that sonic/ultrasonic debridement can result in saving a considerable amount of time, thus being more ergonomic compared to hand instrumentation of periodontal pockets [39]. Therefore, when one considers the demands of clinical skill, time and stamina, the sonic/ultrasonic scaler seems to be the instrument of choice [40].

It has been shown that a single session of root planing can significantly reduce the bacterial population, even without effectively removing all subgingival calculus [41–43]. Areas of limited access – such as inter-proximal areas, furcation lesions, deep vertical periodontal defects, anatomical restrictions on the root – are most likely to exhibit small deposits of residual calculus following instrumentation [44–49], and interestingly surfaces with remaining calculus in the absence of bacterial plaque can be compatible with a clinically acceptable level of gingival wound healing [50–52]. Minute amounts of residual endotoxin detected following root surface debridement were not related to the widely varying amounts of remaining calculus, indicating that bacterial plaque rather than calculus has a greater pathogenic potential [35]. Therefore, the traditional non-surgical periodontal treatment, which comprises of subgingival scaling for the removal of calculus and exposed cementum (as opposed to 'debridement' for the removal of subgingival biofilm), should be questioned in periodontal practice [35, 53]. However, subgingival calculus has the potential to increase the detrimental effect of plaque as it acts as a plaque-retaining factor, and therefore the removal of calculus with overlying plaque is fundamental in the cause-related phase of periodontal therapy.

Timing of Non-Surgical Periodontal Treatment Modalities

Over the years, periodontal treatment has been criticized for its time length, especially in advanced cases, which seems to be comparable to, or even exceeds, that required for a major medical operation. Lately, the time frame of periodontal treatment has received increasing attention in the literature to the extent that it has become central to clinical protocols of non-surgical periodontal treatment. Nowadays, as treatment regimes tend to get shorter, a research question is being posed to clinicians: whether or not the time of full-mouth instrumentation can affect the treatment outcome and in what direction.

The overall time period required to complete periodontal treatment has been suggested in the literature as being highly significant in terms of preventing re-colonization of already treated sites by the remaining untreated pockets and/or from other intraoral niches [54]. It may well be that full-mouth instrumentation completed in a short period of time causes a large-scale bacteremia that may boost the host immune response, hence offering a superior clinical outcome [55]. Shorter treatment regimens have also evident practical benefits in that fewer appointments and less chair time for treatment are required. In addition, the very short protocols of mechanical periodontal treatment by means of ultrasonic debridement combined with thorough plaque control measures have been shown to cause a minimum degree of discomfort to the patient, yet demonstrate similar clinical efficacy with the more stringent therapy of scaling and root planing [56, 57].

Advances in Non-Surgical Periodontal Treatment Protocols

The effectiveness of calculus removal by a single episode of scaling and root planing was evaluated, and compared with that removed by three episodes of hand instrumentation, which seemed to offer no additional clinical benefit [58]. Similarly, an earlier study pointed out that repeated instrumentation of deep periodontal pockets may be of limited value, since a single session of subgingival ultrasonic debridement combined with meticulous plaque control measures was successful in the management of deep sites at single rooted teeth over a 2-year observation period [59]. A major therapeutical concern is that following instrumentation and during the healing phase, subgingival re-colonization may occur. However, the lack of difference in the healing response of periodontal tissues between a single and multiple sessions of instrumentation indicates that re-colonization during a period when sufficient gingival healing has not taken place does not appear to be a major clinical problem in subjects maintaining a high standard of oral hygiene [59].

On this basis, Quirynen et al. [60] introduced the one-stage full-mouth disinfection protocol, and compared the clinical and microbiological effects of this treatment strategy with the more traditional treatment of quadrant scaling and root planing at two-weekly intervals with no adjunctive use of chlorhexidine. In brief, the one-stage protocol included full-mouth scaling and root planing under local anesthesia using hand and ultrasonic instruments, completed over 2 consecutive days within 24 h. Time spent for scaling/root planing one quadrant was approximately 1 h. The disinfection of the oral cavity was undertaken by meticulous application of chlorhexidine products to all intraoral niches, such as periodontal pockets, tongue dorsum, tonsils and oral mucous membranes (chairside, and at home for 2 weeks post-treatment). The authors advocated that the one-stage treatment with the adjunctive use of chlorhexidine prevented subgingival re-colonization. In the following years, from 1995 to 2006, the Leuven research group conducted a series of clinical trials and consistently demonstrated a superior clinical outcome for the one-stage full-mouth disinfection treatment [54, 60–64]. This treatment strategy developed into the clinical protocol of the one-stage full-mouth scaling and root planing with no adjunctive use of antiseptic agents. However, the reported benefits of the one-stage treatment were not verified by other research groups who used hand- and/or power-driven instruments [56, 57, 65–68]. These conflicting findings have generated an ongoing debate in the literature as to whether the full-mouth clinical protocol should be the treatment of choice. The overview of the relevant studies to this treatment protocol and their main findings are depicted in table 1.

Apatzidou and Kinane [65] randomized 40 patients into two treatment groups (one-day full-mouth root planing versus quadrant-wise root planing on a bi-weekly basis) and followed these subjects over 6 months. The full-mouth group received scaling and root planing in the upper and lower quadrants on the right using hand and ultrasonic instruments under local anesthesia in a morning session, while the remaining two quadrants on the left were instrumented in the afternoon session of the same day and in the same manner as the previously treated segments. Time spent for each treatment session was approximately 2 h, with a 1-hour break in between. It should be borne in mind that there was no use of antiseptic products, and although the

Table 1. Summary of the clinical studies regarding advances in non-surgical treatment protocols

Authors	Treatment modality	Description of participants	Duration of the study, months	Main findings
Quirynen et al. [60]	FMD followed by CHX application for 2 weeks vs. biweekly Q-SRP	FMD n = 5 Q-SRP n = 5 advanced chronic periodontitis patients	2	significantly greater pocket depth reduction for FMD at deep pockets only
Vandekerckhove et al. [61]	FMD followed by CHX application for 2 weeks vs. biweekly Q-SRP	FMD n = 5 Q-SRP n = 5 advanced chronic periodontitis patients	8	similar findings as the 1995 study
Bollen et al. [62]	FMD followed by CHX application for 2 months vs. biweekly Q-SRP	FMD n = 8 Q-SRP n = 8 advanced chronic periodontitis patients	4	extended use of CHX gave additional improvements for FMD treatment
Mongardini et al. [63]	FMD followed by CHX application for 2 months vs. biweekly Q-SRP	FMD n = 20 Q-SRP n = 20 16 generalized aggressive and 24 advanced chronic periodontitis patients	8	FMD more efficacious than Q-SRP, even in aggressive forms of periodontitis
Quirynen et al. [54]	FMD followed by CHX application for 2 months vs. FM-SRP within 24 h vs. biweekly Q-SRP	FMD n = 12 Q-SRP n = 12 FM-SRP n = 12 advanced chronic periodontitis patients; FMD and Q-SRP groups were recruited from the 1999 study	8	FMD and FMRP equally efficacious, but more favorable than Q-SRP
Apatzidou and Kinane [55]	same-day FM-SRP vs. biweekly Q-SRP	Q-SRP n = 20 FM-SRP n = 20 moderately advanced chronic periodontitis patients	6	both treatments equally efficacious
Koshy et al. [56]	weekly QMD vs. single-visit FMD_{water} vs. single-visit $FMD_{povidone}$, followed by CHX 0.05% for 1 month	FMD n = 12 FMD+water n = 12 FMD+povidone n =12 moderately advanced chronic periodontitis patients	6	no differences in pocket depth reductions between treatment groups

Table 1. Continued

Authors	Treatment modality	Description of participants	Duration of the study, months	Main findings
Wennstrom et al. [57]	1-hour FMD_{water} vs. weekly Q-SRP	Q-SRP n = 21 FMD+water n = 20 moderately advanced chronic periodontitis patients	6	1-hour FMD is a justified initial treatment approach
Jervøe-Storm et al. [66]	FM-SRP within 24 h vs. biweekly Q-SRP	Q-SRP n = 10 FM-SRP n =10 moderately advanced chronic periodontitis patients	6	equally favorable results following both treatments
Zanatta et al. [67]	45-min FMD_{PIG} vs. 45-min FMD_{NaCl} vs. weekly Q-SRP	Q-SRP n = 13, FMD+PIG n = 15 FMD + NaCl n = 12 moderately advanced chronic periodontitis patients	3	equally favorable results following all treatments
Quirynen et al. [64]	FMD followed by CHX mouth rinse for 2 months or AmF/SnF_2 for 2 months, or CHX for 2 months and AmF/SnF_2 for 6 months vs. FM-SRP within 24 h vs. biweekly Q-SRP	FM-CHX n =14 FM-F n = 14 FM-CHX+F n = 14 Q-SRP n = 15 FM-SRP = 14 moderately advanced chronic periodontitis patients	8	significantly greater pocket depth reduction for the CHX groups compared to the Q-SRP group, but no differences between any other group
Swierkot et al. [68]	FMD followed by CHX application for 2 weeks vs. FM-SRP within 24 hours vs. weekly Q-SRP	FMD n = 9, Q-SRP n = 7 FM-SRP n = 9 chronic periodontitis patients	8	interim clinical and PCR differences between groups, which disappeared at 8 months

CHX = Chlorhexidine; FMD = one-stage full-mouth disinfection; Q-SRP = quadrant scaling and root planing (no use of antiseptics); FM-SRP = full-mouth root planing (no use of antiseptics); FM-CHX = one-stage full-mouth disinfection (followed by chlorhexidine 0.2% mouth rinse); FM-F = one-stage full-mouth disinfection (followed by AmF/SnF_2 mouth rinse); FM-CHX+F = one-stage full-mouth disinfection (followed by chlorhexidine 0.2% and AmF/SnF_2 mouth rinses); QMD = quadrant supra- and subgingival ultrasonic debridement with water as irrigant (followed by no antiseptics); FMD_{water} = full-mouth supra- and subgingival ultrasonic debridement with water as irrigant, followed by no antiseptics; $FMD_{povidone}$ = full-mouth supra- and subgingival ultrasonic debridement with povidone iodine as irrigant, followed by the use of chlorhexidine 0.05% mouth rinse; FMD_{PIG} = full-mouth supra- and subgingival ultrasonic debridement with 0.5% pvp-iodine as irrigant; FMD_{NaCl} = full-mouth supra- and subgingival ultrasonic debridement with 0.9% NaCl as irrigant.

1-day group was mechanically treated on the same day, these subjects were still seen every second week in order to receive the same amount of oral hygiene instructions and motivation as the quadrant-wise group. This resulted in similar plaque indices between the treatment groups throughout the study, which contrasts with previous findings that showed higher plaque indices for the one-stage full-mouth disinfection group after the first month, possibly due to a lack of frequent sessions of oral hygiene reinforcement [61].

Koshy et al. [56] re-analyzed the effects of full-mouth and quadrant-wise treatment by using ultrasonics rather than hand instrumentation. Their protocol, briefly stated, was to complete full-mouth ultrasonic debridement in ≤2.5 h with either povidone iodine or water as irrigants of the ultrasonic device, and compare the clinical outcome and PCR findings of this treatment with the quadrant-wise ultrasonic debridement at weekly intervals. Their results demonstrated that the single-visit debridement was not more advantageous over the partial treatment, except that less time was required overall to complete treatment in one visit than over four consecutive sessions.

More interestingly, Wennström et al. [57] reduced the time frame of full-mouth periodontal therapy in moderately advanced chronic periodontitis patients into 1 h with the use of ultrasonics solely under local anesthesia, if requested by the patient. Although the 1-hour treatment resulted in a smaller percentage of 'closed' pockets (probing pocket depth <5 mm) compared with the classical treatment of scaling and root planing per quadrant on a weekly basis, the efficiency of the initial treatment phase (baseline to 3 months), defined as the time spent for instrumentation divided by the number of closed pockets, was significantly higher for the full-mouth ultrasonic debridement. In more detail, 58% and 3.3 min were the percentage of closed pockets and the efficiency of the initial treatment at 3 months, respectively, for the 1-hour treatment, as opposed to 66% and 8.8 min for the quadrant-wise treatment. Emphasis should be placed upon the fact that in this study oral hygiene instructions were given on three occasions in a similar manner for both treatment groups. In a previous study of the same research group, a maximum of 45 min was allowed for full-mouth debridement of chronic periodontitis patients followed by locally delivered antimicrobials in pockets with an initial pocket depth ≥5 mm [69]. Summarizing the results of these studies, the 3-month re-assessment demonstrated a similar clinical outcome between the traditional quadrant-wise treatment and the simplified compact treatment of 1 hour or 45 min, which in combination with careful oral hygiene measures has been suggested as a justifiable initial treatment approach for the chronic periodontitis patient [57, 69]. However, of great importance is the fact that although at 3 months periodontal sites with probing pocket depth ≥5 mm received additional instrumentation with no time limitation, approximately 50% of the initially deep pockets (≥7 mm) still remained as non-successful sites at 6 months [57]. Therefore, decision-making as to whether repeated instrumentation or other treatment (such as surgical or antibiotic treatments) would be beneficial for these poorly responding sites is critical and should be justified in each case.

Nevertheless, one has to take into serious consideration that periodontitis is a chronic multifactorial inflammatory disease process, which requires the long-term commitment of both the patient and the therapist in order to achieve long-term periodontal stability. Although the time frame of a full-mouth treatment can range from a few hours over multiple visits to less than 1 h in a single visit, the therapist should monitor periodontitis patients closely during the initial phase of treatment, especially in advanced cases where the oral hygiene practices and patient compliance are important elements of a successful periodontal therapy [70]. The reported data underline the significance of home care measures in adjunction to professionally performed mechanical removal of the biofilm in the resolution and prevention of periodontal disease. Scaling and root planing is a technically demanding and time-consuming therapeutical procedure. Considering that approximately 1 h is required for root planing each quadrant, the one-stage full-mouth scaling and root planing offers no additional economic advantages, except for less time and cost of transit and reduced use of materials in the clinic, since two visits are needed instead of four compared with the classical quadrant-wise treatment. In addition to the considerable amount of time required for root planing is the extra chair time needed for the follow-up visits in the immediate healing period to optimize the oral hygiene status of chronic periodontitis patients. This position partially contrasts with the conclusions of the Leuven group regarding the economic advantage of the one-stage treatment [71] and welcomes the practical value of the simplified one-visit full-mouth ultrasonic debridement [72].

Host Response Changes Following Non-Surgical Periodontal Therapy

There is ample evidence in the literature to support the impact of periodontal therapy on the local and systemic host immune responses that protect the host tissues from the microbial challenge. Research data suggest that antibody levels decrease following treatment, probably due to a reduced microbial load [73–79, 55], whereas other studies demonstrate increased antibody levels post-treatment that are possibly explained by inoculation of bacteria into the host tissues and boost of the immune response [80, 81]. These conflicting data could be attributed to differences in laboratory protocols and clinical study designs, in particular the time of testing following therapy. However, the majority of the literature agrees that increased antibody avidity post-treatment indicates maturation of the immune system, and therefore successful treatment [82, 83].

Quirynen et al. [54] observed that 7 out of 11 patients, whose body temperature rose above 37°C after the second day of the one-stage treatment had an overall pocket depth reduction ≥3.5 mm, whereas this was noted for only 4 of the remaining 13 patients, who did not have an increase in temperature. The observation that patients with a rise in body temperature the evening after the second day of the full-mouth treatment had the more impressive clinical improvements was considered by the authors to be due to an increased immunological reaction, which might explain the enhanced clinical outcome noted after this treatment strategy.

Apatzidou et al. [55] initially aimed to investigate whether the full-mouth treatment strategy would augment the host response, and hence offer a superior clinical outcome, but their results failed to support this hypothesis and indeed provide evidence for the superiority of this treatment over quadrant scaling. This study collected frequent blood samples from patients (biweekly intervals) over a longer period (overall 6 months) to determine the dynamics of the immune response following the compact clinical protocol versus the traditional quadrant-wise treatment. The data showed that although full-mouth instrumentation was completed in a shorter period than previously described (< 12h on the same day rather than 24h over two consecutive days) [60], no significant differences in the IgG antibody titres against five putative periodontal pathogens were detected compared to the standard quadrant root planing at biweekly intervals over a period of 6 months. It may well be that fortnightly sessions of quadrant root planing result in a similar host immune response as same-day or 24-hour full-mouth treatment, even though the former treatment approach would be expected to have a greater potential for repeated inoculation of bacteria into the host tissues, while the latter treatment would be considered to cause a larger immunological reaction, due to the one-stage introduction of bacteria and their products into the underlining tissues. Although IgG antibody titres were similar for both treatment groups, a significantly greater reduction in the antibody levels to *Prevotella intermedia* and *Treponema denticola* was noted between baseline and 6 weeks after the completion of instrumentation (initial reassessment) for the full-mouth-treated group compared with the quadrant-wise group. When the longitudinal changes in the serum antibody response were considered for each treatment group over the 6-month duration of the study, significant reductions in the IgG antibody levels to *P. intermedia* and *T. denticola* were seen between baseline and the initial reassessment for the full-mouth group, but this was not the case for the quadrant-wise group. These findings suggest that same-day full-mouth treatment seems to have a stronger short-term effect on the systemic antibody response compared to the classic therapy of quadrant root planing, which complies with more recent data [84], and may reflect the one-stage reduction of the microbial burden. However, the clinical significance of this finding is difficult to assess.

A distinct trend was noted for the binding ability or avidity of antibodies to increase at 6 months, but this was not found to be statistically significant for the majority of test microorganisms, reflecting marked inter-individual variation [55]. Nevertheless, significant short-term increases in antibody avidity to most test bacteria were noted for both treatment strategies, which are in agreement with other data [84]. It should be noted that there were no significant differences in antibody avidity between the full-mouth and the quadant-wise treatment at any time point [55, 84]. In conclusion, short-term perturbations of antibody responses occur during the active phase of periodontal therapy, but show great variation across subjects, and are therefore difficult to meaningfully interpret [55].

Subgingival Microflora Changes Induced by Non-Surgical Periodontal Treatment

Periodontal treatment aims to remove plaque and calculus in order to reduce quantities of subgingival bacteria below a threshold level capable of causing clinical inflammation and loss of periodontal tissues [85]. Several studies sought to determine the beneficial effects of scaling and root planing on both clinical and microbiological parameters and have consistently reported shifts to a less pathogenic subgingival microflora coupled with improvements in the periodontal status [86–90]. A thorough review of the literature documented that subgingival scaling by and large decreased the population of gram-negative microbes, while at the same time increased the populations of gram-positive rods and cocci; a change that is usually compatible with periodontal health [40]. Nevertheless, it is unlikely that scaling and root planing can eradicate and permanently clear all microorganisms from a site [89], signifying that the mere presence of a suspected pathogen in a site is not indicative of disease presence [91]. Radical alterations in the subgingival microflora may not be necessary or even desirable, since the host-parasite equilibrium may dramatically change for the detriment of the host [89]. Optimal treatment outcome is guaranteed by thorough plaque control, which ensues from highly motivated and skilful patients in oral hygiene practises for the long-term maintenance of plaque-free periodontal tissues [92–94].

The analysis of the microbiological parameters following non-surgical instrumentation demonstrated that when identification methods based on molecular biology techniques were employed to determine the quantity or the detection frequency of putative periodontal pathogens in subgingival plaque, no significant differences were found between the full-mouth approach and the quadrant-wise treatment [95, 56, 96]. These data contrast findings of De Soete et al. [97], who utilized DNA-DNA hybridization techniques and reported greater reductions in the 'red' and 'orange' complex bacteria for the full-mouth treatment modality. Similarly, further studies from the University of Leuven using conventional culturing techniques [54, 62, 98, 99] disagree with reports that come from the University of Glasgow [95], Tokyo [56] and Bonn [96], which makes the interpretation of this inconsistency rather difficult.

In the study by Apatzidou et al. [95], although the time interval between initiation and completion of full-mouth scaling and root planing was lessened, and therefore the chances for bacterial re-colonization were also reduced, no significant microbiological or clinical differences were found between the two treatment groups 6 months post-scaling to verify subgingival re-colonization. A more recent study demonstrated that scaling and root planing completed in a single visit versus multiple sessions resulted in less re-colonization of *T. denticola* at a single site in the maxillary right quadrant at 3 months [100]. Re-colonization was designated as when a site was scored positive for a species by PCR before treatment, turned negative after treatment, and became positive again thereafter. The authors concluded that the full-mouth treatment may be more effective in preventing re-colonization of periodontal lesions. However, the interpretation of this finding should be taken with care, due to the fact that both treatment approaches were found to be equally efficacious at 3 months, based on full-mouth clinical measurements and PCR-data of five species in subgingival plaque collected from four sites per patient.

In conclusion, the fact that the results obtained from a variety of microbiological identification methods are difficult to directly compare led Lang et al. [101] to conduct a meta-analysis of microbiological results of the full-mouth treatment concept in a systematic review undertaken for the 6th European Workshop on Periodontology. The authors concluded that if re-colonization of the already scaled sites was prevented by the full-mouth treatment, then the study of Jervøe-Storm et al. [96] – which examined changes in the quantity of a number of putative periodontal pathogens in subgingival plaque of the two deepest pockets in the first quadrant by elaborated laboratory methods before treatment and after 1 day and 1, 2, 4, 8, 12 and 24 weeks – should have validated this hypothesis, but failed to do so.

Conclusions

A critical question in the full-mouth disinfection protocol relates to which element of this treatment strategy contributes mostly to the enhanced treatment outcome reported in the literature. One speculation is that the extensive and multiple applications of antiseptic products for a prolonged period of time have resulted in sustained bacterial plaque control in all intraoral niches [62]. Evidence shows that the use of chlorhexidine does not augment the beneficial outcome of periodontal therapy [102–104], and when this does occur, it seems to be a short-term effect rather than a long-lasting event [105, 106]. The addition of antiseptic agents to coolants or irrigants of ultrasonic devices does not offer additional clinical improvement [107]. In brief, there is a general agreement in the literature that the clinical trials performed without meticulous plaque control gave misleadingly promising results for the adjunctive use of chlorhexidine. However, it should be borne in mind that in the quoted studies chlorhexidine was used as a single measure of disinfection, in contrast to the treatment protocol of multiple chlorhexidine applications used in the Leuven studies [54, 60–64]. However, one study from this research group questioned the role of the antiseptics in the one-stage full-mouth treatment, since full-mouth scaling and root planing with no antiseptics was equally efficacious with the one-stage full-mouth disinfection, and both treatments were superior to quadrant scaling and root planing with no antiseptics [54].

On the other hand, it could be that the mechanical treatment per se, which was completed within a relatively short period of time in the full dentition, decreased the microbial burden in one stage, and thus lessened the chances for subgingival re-colonization [54]. Literature on the time frame of non-surgical periodontal treatment points out that whether full-mouth instrumentation with no antiseptics is completed in 1 h [57], ≤2.5 h [56], 1 day [65], 2 consecutive days [54] or 6 weeks (traditional protocol of quadrant scaling and root planing on a biweekly basis), it has no significant impact on the treatment outcome.

Lastly, the benefits of the full-mouth treatment protocol could be partially due to the antiseptics and partially to the completion of treatment in a shorter time period [64]. It may also be that less thorough and extensive subgingival instrumentation is required for the treatment of moderately advanced chronic periodontitis patients, when locally delivered antimicrobials are used as an adjunct in the sites ≥5 mm [69]. In addition, the full-mouth treatment approach using ultrasonic debridement, in combination with systemic antibiotics [Mombelli, pp. 133–148], may prove to be the optimal protocol especially for the management of specific patient groups [108]. In this case, reduced treatment time is considered a tangible benefit for the patient and this simplified one-visit treatment may become a significant paradigm shift in periodontal practice [72].

In summary, no significant differences were found between scaling and root planing with the sole use of hand instruments versus ultrasonic debridement, nor between quadrant-wise approaches versus full-mouth approaches. There is a lack of convincing and consistent evidence in the literature that the full-mouth disinfection approach is superior to full-mouth scaling and root planing and/or the traditional quadrant-wise therapy [109]. Thus, we can conclude that all three treatment options are equally recommended for debridement in the cause-related phase of periodontal therapy in patients with chronic periodontitis [101]. The mechanical removal of the microbial biofilm and the establishment of meticulous plaque control measures comprise the key elements for the success of non-surgical periodontal treatment. The clinician should select the treatment modality based on practical considerations related to time constraints, clinical workload and patient preference [65, 107].

References

1 Socransky SS, Haffajee AD: Dental biofilms: difficult therapeutic targets. Periodontol 2000 2002;28: 12–55.

2 Listgarten MA, Lindhe J, Hellden L: Effect of tetracycline and/or scaling on human periodontal disease: clinical, microbiological, and histological observations. J Clin Periodontol 1978;5:246–271.

3 Kho P, Smale FC, Hardi JM: The effect of supragingival plaque control on the subgingival microflora. J Clin Periodontol 1985;12:676–686.

4 Beltrami M, Bickel M, Baehni PC: The effect of supragingival plaque control on the composition of the subgingival microflora in human periodontitis. J Clin Periodontol 1987;14:161–164.

5 Loos B, Claffey N, Crigger M: Effects of oral hygiene measures on clinical and microbiological parameters of periodontal disease. J Clin Periodontol 1988; 15:211–216.

6 Lavanchy DL, Bickel M, Baehni PC: The effect of plaque control after scaling and root planing on the subgingival microflora in human periodontitis. J Clin Periodontol 1987;14:295–299.

7 Magnusson I, Lindhe J, Yoneyama T, Liljenberg B: Recolonization of a subgingival microbiota following scaling in deep pockets. J Clin Periodontol 1984; 11:193–207.

8 Müller HP, Hartmann J, Flores-de-Jacoby L: Clinical alterations in relation to the morphological composition of the subgingival microflora following scaling and root planing. J Clin Periodontol 1986;13:825–832.
9 Ximénez-Fyvie LA, Haffajee AD, Som S, Thompson M, Torresyap G, Socransky SS: The effect of repeated professional supragingival plaque removal on the composition of the supra- and subgingival microbiota. J Clin Periodontol 2000;27:637–647.
10 Haffajee AD, Smith C, Torresyap G, Thompson M, Guerrero D, Socransky SS: Efficacy of manual and powered toothbrushes (II). Effect on microbiological parameters. J Clin Periodontol 2001;28:947–954.
11 Smulow JB, Turesky SS, Hill RG: The effect of supragingival plaque removal on anaerobic bacteria deep periodontal pockets. J Am Dent Assoc 1983;107:737–742.
12 Dahlén G, Lindhe J, Sato K, Hanamura H, Okamoto H: The effect of supragingival plaque control on the subgingival microbiota in subjects with periodontal disease. J Clin Periodontol 1992;19:802–809.
13 McNabb H, Mombelli A, Lang NP: Supragingival cleaning 3 times a week: the microbiological effects in moderately deep pockets. J Clin Periodontol 1992;19:348–356.
14 Katsanoulas T, Reneè I, Attström R: The effect of supragingival plaque control on the composition of the subgingival flora in periodontal pockets. J Clin Periodontol 1992;19:760–765.
15 Hellström MK, Ramberg P, Krok L, Lindhe J: The effect of supragingival plaque control on the subgingival microflora in human periodontitis. J Clin Periodontol 1996;23:934–940.
16 Mousquès T, Listgarten MA, Phillips RW: Effect of scaling and root planing on the composition of the human subgingival microbial flora. J Periodontal Res 1980;15:144–151.
17 Aleo JJ, De Renzis FA, Farber PA: In vitro attachment of human gingival fibroblasts to root surfaces. J Periodontol 1975;46:639–645.
18 Jones WA, O'Leary TJ: The effectiveness of in vivo root planing in removing bacterial endotoxin from the roots of periodontally involved teeth. J Periodontol 1978;49:337–342.
19 Nishimine D, O'Leary TJ: Hand instrumentation versus ultrasonics in the removal of endotoxins from root surfaces. J Periodontol 1979;50:345–349.
20 Hatfield CG, Baumhammers A: Cytotoxic effects of periodontally involved surfaces of human teeth. Arch Oral Biol 1971;16:465–468.
21 Aleo JJ, De Renzis FA, Farber PA, Varboncoeur AP: The presence and biologic activity of cementum-bound endotoxin. J Periodontol 1974;45:672–675.
22 Moore J, Wilson M, Kieser JB: The distribution of bacterial lipopolysaccharide (endotoxin) in relation to periodontally involved root surfaces. J Clin Periodontol 1986;13:748–751.
23 Hughes FJ, Auger DW, Smales FC: Investigation of the distribution of cementum-associated lipopolysaccharides in periodontal disease by scanning electron microscope immunohistochemistry. J Periodontal Res 1988;23:100–106.
24 Cadosch J, Zimmermann U, Ruppert M, Guindy J, Case D, Zappa U: Root surface debridement and endotoxin removal. J Periodontal Res 2003;38:229–236.
25 Nyman S, Westfelt E, Sarhed G, Karring T: Role of diseased root cementum in healing following treatment of periodontal disease: a clinical study. J Clin Periodontol 1988;15:464–468.
26 Wirthlin MR, Hancock EB, Gaugler RW: Regeneration and repair after biologic treatment of root surfaces in monkeys. J Periodontol 1981;52:729–735.
27 Nyman S, Sarhed G, Ericsson I, Gottlow J, Karring T: Role of diseased root cementum in healing following treatment of periodontal disease: an experimental study in the dog. J Periodontal Res 1986;21:496–503.
28 Blomlöf L, Lindskog S, Appelgren R, Jonsson B, Weintraub A, Hammarström L: New attachment in monkeys with experimental periodontitis with and without removal of the cementum. J Clin Periodontol 1987;14:136–143.
29 Busslinger A, Lampe K, Beuchat M, Lehmann B: A comparative in vitro study of a magnetostrictive and a piezoelectric ultrasonic scaling instrument. J Clin Periodontol 2001;28:642–649.
30 Schmidlin PR, Beuchat M, Busslinger A, Lehmann B, Lutz F: Tooth substance loss resulting from mechanical, sonic and ultrasonic root instrumentation assessed by liquid scintillation. J Clin Periodontol 2001;28:1058–1066.
31 Walsh TF, Waite IM: A comparison of postsurgical healing following debridement by ultrasonic or hand instruments. J Periodontol 1978;49:201–205.
32 Torfason T, Kiger R, Selvig KA, Egelberg J: Clinical improvement of gingival conditions following ultrasonic versus hand instrumentation of periodontal pockets. J Clin Periodontol 1979;6:165–176.
33 Badersten A, Nilveus R, Egelberg J: Effect of nonsurgical periodontal therapy. II. Severely advanced periodontitis. J Clin Periodontol 1984;11:63–76.
34 Eschler BM, Rapley JW: Mechanical and chemical root preparation in vitro: efficiency of plaque and calculus removal. J Periodontol 1991;62:755–760.

35 Chiew SY, Wilson M, Davies EH, Kieser JB: Assessment of ultrasonic debridement of calculus-associated periodontally-involved root surfaces by the limulus amoebocyte lysate assay: an in vitro study. J Clin Periodontol 1991;18:240–244.
36 Copulos TA, Low SB, Walker CB, Trebilcock YY, Hefti AF: Comparative analysis between a modified ultrasonic tip and hand instruments on clinical parameters of periodontal disease. J Periodontol 1993;64:694–700.
37 Drisko CL, Cochran DL, Blieden T, Bouwsma OJ, Cohen RE, Damoulis P, Fine JB, Greenstein G, Hinrichs J, Somerman MJ, Iacono V, Genco RJ, Science and Therapy Committee of the American Academy of Periodontology Research: Position paper: sonic and ultrasonic scalers in periodontics. Research, Science and Therapy Committee of the American Academy of Periodontology. J Periodontol 2000;71:1792–1801.
38 Walmsley AD, Lea SC, Landini G, Moses AJ: Advances in power driven pocket/root instrumentation. J Clin Periodontol 2008;35(suppl):22–28.
39 Tunkel J, Heinecke A, Flemmig TF: A systematic review of efficacy of machine-driven and manual subgingival debridement in the treatment of chronic periodontitis. J Clin Periodontol 2002;29(suppl 3): 72–81.
40 Cobb CM: Clinical significance of non-surgical periodontal therapy: an evidence-based perspective of scaling and root planing. J Clin Periodontol 2002; 29(suppl 2):6–16.
41 Waerhaug J: Healing of the dento-epithelial junction following subgingival plaque control. II. As observed on extracted teeth. J Periodontol 1978;49: 119–134.
42 Breininger DR, O'Leary TJ, Blumenshine RV: Comparative effectiveness of ultrasonic and hand scaling for the removal of subgingival plaque and calculus. J Periodontol 1987;58:9–18.
43 Sherman PR, Hutchens LH Jr, Jewson LG, Moriarty JM, Greco GW, McFall WT Jr: The effectiveness of subgingival scaling and root planning. I. Clinical detection of residual calculus. J Periodontol 1990; 61:3–8.
44 Caffesse RG, Sweeney PL, Smith BA: Scaling and root planing with and without periodontal flap surgery. J Clin Periodontol 1986;13:205–210.
45 Buchanan SA, Robertson PB: Calculus removal by scaling/root planing with and without surgical access. J Periodontol 1987;58:159–163.
46 Brayer WK, Mellonig JT, Dunlap RM, Marinak KW, Carson RE: Scaling and root planing effectiveness: the effect of root surface access and operator experience. J Periodontol 1989;60:67–72.
47 Fleischer HC, Mellonig JT, Brayer WK, Gray JL, Barnett JD: Scaling and root planing efficacy in multirooted teeth. J Periodontol 1989;60:402–409.
48 Wylam JM, Mealey BL, Mills MP, Waldrop TC, Moskowicz DC: The clinical effectiveness of open versus closed scaling and root planing on multirooted teeth. J Periodontol 1993;64:1023–1028.
49 Kocher T, Gutsche C, Plagmann HC: Instrumentation of furcation with modified sonic scaler inserts: study on manikins, part I. J Clin Periodontol 1998;25:388–393.
50 Listgarten MA, Ellegaard B: Electron microscopic evidence of a cellular attachment between junctional epithelium and dental calculus. J Periodontal Res 1973;8:143–150.
51 Fujikawa K, O'Leary TJ, Kafrawy AH: The effect of retained subgingival calculus on healing after flap surgery. J Periodontol 1988;59:170–175.
52 Biagini G, Checchi L, Pelliccioni GA, Solmi R: In vitro growth of periodontal fibroblasts on treated cementum. Quintessence Int 1992;23:335–340.
53 Mombelli A, Nyman S, Brägger U, Wennström J, Lang NP: Clinical and microbiological changes associated with an altered subgingival environment induced by periodontal pocket reduction. J Clin Periodontol 1995;22:780–787.
54 Quirynen M, Mongardini C, de Soete M, Pauwels M, Coucke W, van Eldere J, van Steenberghe D: The role of chlorhexidine in the one-stage full-mouth disinfection treatment of patients with advanced adult periodontitis: long-term clinical and microbiological observations. J Clin Periodontol 2000;27: 578–589.
55 Apatzidou DA, Kinane DF: Quadrant root planing versus same-day full-mouth root planing. III. Dynamics of the immune response. J Clin Periodontol 2004;31:152–159.
56 Koshy G, Kawashima Y, Kiji M, Nitta H, Umeda M, Nagasawa T, Ishikawa I: Effects of single-visit full-mouth ultrasonic debridement versus quadrant-wise ultrasonic debridement. J Clin Periodontol 2005;32:734–743.
57 Wennström JL, Tomasi C, Bertelle A, Dellasega E: Full-mouth ultrasonic debridement versus quadrant scaling and root planing as an initial approach in the treatment of chronic periodontitis. J Clin Periodontol 2005;32:851–859.
58 Anderson GB, Palmer JA, Bye FL, Smith BA, Caffesse RG: Effectiveness of subgingival scaling and root planing: single versus multiple episodes of instrumentation. J Periodontol 1996;67:367–373.
59 Badersten A, Nilveus R, Egelberg J: Effect of nonsurgical periodontal therapy. III. Single versus repeated instrumentation. J Clin Periodontol 1984; 11:114–124.

60 Quirynen M, Bollen CM, Vandekerckhove BN, Dekeyser C, Papaioannou W, Eyssen H: Full- vs. partial-mouth disinfection in the treatment of periodontal infections: short-term clinical and microbiological observations. J Dent Res 1995;74:1459–1467.
61 Vandekerckhove BN, Bollen CM, Dekeyser C, Darius P, Quirynen M: Full- versus partial-mouth disinfection in the treatment of periodontal infections: long-term clinical observations of a pilot study. J Periodontol 1996;67:1251–1259.
62 Bollen CM, Mongardini C, Papaioannou W, Van Steenberghe D, Quirynen M: The effect of a one-stage full-mouth disinfection on different intra-oral niches: clinical and microbiological observations. J Clin Periodontol 1998;25:56–66.
63 Mongardini C, van Steenberghe D, Dekeyser C, Quirynen M: One stage full- versus partial-mouth disinfection in the treatment of chronic adult or generalized early-onset periodontitis. I. Long-term clinical observations. J Periodontol 1999;70:632–645.
64 Quirynen M, De Soete M, Boschmans G, Pauwels M, Coucke W, Teughels W, van Steenberghe D: Benefit of one-stage full-mouth disinfection is explained by disinfection and root planing within 24 hours: a randomized controlled trial. J Clin Periodontol 2006;33:639–647.
65 Apatzidou DA, Kinane DF: Quadrant root planing versus same-day full-mouth root planing. I. Clinical findings. J Clin Periodontol 2004;31:132–140.
66 Jervøe-Storm PM, Semaan E, AlAhdab H, Engel S, Fimmers R, Jepsen S: Clinical outcomes of quadrant root planing versus full-mouth root planing. J Clin Periodontol 2006;33:209–215.
67 Zanatta GM, Bittencourt S, Nociti FH Jr, Sallum EA, Sallum AW, Casati MZ: Periodontal debridement with povidone-iodine in periodontal treatment: short-term clinical and biochemical observations. J Periodontol 2006;77:498–505.
68 Swierkot K, Nonnenmacher CI, Mutters R, Flores-de-Jacoby L, Mengel R: One-stage full-mouth disinfection versus quadrant and full-mouth root planing. J Clin Periodontol 2009;36:240–249.
69 Wennström JL, Newman HN, MacNeill SR, Killoy WJ, Griffiths GS, Gillam DG, Krok L, Needleman IG, Weiss G, Garrett S: Utilisation of locally delivered doxycycline in non-surgical treatment of chronic periodontitis: a comparative multi-centre trial of 2 treatment approaches. J Clin Periodontol 2001;28:753–761.
70 Apatzidou DA: One stage full-mouth disinfection-treatment of choice? J Clin Periodontol 2006;33:942–943.
71 Quirynen M, Teughels W, van Steenberghe D: Impact of antiseptics on one-stage, full-mouth disinfection. J Clin Periodontol 2006;33:49–52.
72 Kinane DF: Single-visit, full-mouth ultrasonic debridement: a paradigm shift in periodontal therapy? J Clin Periodontol 2005;32:732–733.
73 Tolo K, Schenck K, Johansen JR: Activity of human serum immunoglobulins to seven anaerobic oral bacteria before and after periodontal treatment. J Periodontal Res 1982;17:481–483.
74 Naito Y, Okuda K, Takazoe I, Watanabe H, Ishikawa I: The relationship between serum IgG levels to subgingival gram-negative bacteria and degree of periodontal destruction. J Dent Res 1985;64:1306–1310.
75 Mouton C, Desclauriers M, Allard H, Bouchard M: Serum antibodies to *Bacteroides gingivalis* in periodontitis: a longitudinal study. J Periodontal Res 1987;22:426–430.
76 Aukhil I, Lopatin DE, Syed SA, Morrison EC, Kowalski CJ: The effects of periodontal therapy on serum antibody (IgG) levels to plaque microorganisms. J Clin Periodontol 1988;15:544–550.
77 Murray PA, Burstein DA, Winkler JR: Antibodies to *Bacteroides gingivalis* in patients with treated and untreated periodontal disease. J Periodontol 1989;60:96–103.
78 Johnson V, Johnson BD, Sims TJ, Whitney CW, Moncla BJ, Engel LD, Page RC: Effects of treatment on antibody titer to *Porphyromonas gingivalis* in gingival crevicular fluid of patients with rapidly progressive periodontitis. J Periodontol 1993;64:559–565.
79 Horibe M, Watanabe H, Ishikawa I: Effect of periodontal treatments on serum IgG antibody titers against periodontopathic bacteria. J Clin Periodontol 1995;22:510–515.
80 Ebersole JL, Taubman MA, Smith DJ, Haffajee AD: Effect of subgingival scaling on systemic antibody responses to oral microorganisms. Infect Immun 1985;48:534–539.
81 Sjöström K, Ou J, Whitney C, Johnson B, Darveau R, Engel D, Page RC: Effect of treatment on titer, function, and antigen recognition of serum antibodies to *Actinobacillus actinomycetemcomitans* in patients with rapidly progressive periodontitis. Infect Immun 1994;62:145–151.
82 Chen HA, Johnson BD, Sims TJ, Darveau RP, Moncla BJ, Whitney CW, Engel D, Page RC: Humoral immune responses to *Porphyromonas gingivalis* before and following therapy in rapidly progressive periodontitis patients. J Periodontol 1991;62:781–791.

83 Mooney J, Adonogianaki E, Riggio MP, Takahashi K, Haerian A, Kinane DF: Initial serum antibody titer to *Porphyromonas gingivalis* influences development of antibody avidity and success of therapy for chronic periodontitis. Infect Immun 1995;63: 3411–3416.

84 Wang D, Koshy G, Nagasawa T, Kawashima Y, Kiji M, Nitta H, Oda S, Ishikawa I: Antibody response after single-visit full-mouth ultrasonic debridement versus quadrant-wise therapy. J Clin Periodontol 2006;33:632–638.

85 Research, Science and Therapy Committee of the American Academy of Periodontology: Treatment of plaque-induced gingivitis, chronic periodontitis, and other clinical conditions. J Periodontol 2001;72: 1790–1800.

86 Slots J, Mashimo P, Levine MJ, Genco RJ: Periodontal therapy in humans. I. Microbiological and clinical effects of a single course of periodontal scaling and root planing, and of adjunctive tetracycline therapy. J Periodontol 1979;50:495–509.

87 Renvert S, Wikström M, Dahlén G, Slots J, Egelberg J: Effect of root debridement on the elimination of *Actinobacillus actinomycetemcomitans* and *Bacteroides gingivalis* from periodontal pockets. J Clin Periodontol 1990;17:345–350.

88 Pedrazzoli V, Kilian M, Karring T, Kirkegaard E: Effect of surgical and non-surgical periodontal treatment on periodontal status and subgingival microbiota. J Clin Periodontol 1991;18:598–604.

89 Haffajee AD, Cugini MA, Dibart S, Smith C, Kent RL Jr, Socransky SS: The effect of SRP on the clinical and microbiological parameters of periodontal diseases. J Clin Periodontol 1997;24:324–334.

90 Darby IB, Mooney J, Kinane DF: Changes in subgingival microflora and humoral immune response following periodontal therapy. J Clin Periodontol 2001;28:796–805.

91 Doungudomdacha S, Rawlinson A, Walsh TF, Douglas CW: Effect of non-surgical periodontal treatment on clinical parameters and the numbers of *Porphyromonas gingivalis*, *Prevotella intermedia* and *Actinobacillus actinomycetemcomitans* at adult periodontitis sites. J Clin Periodontol 2001;28:437–445.

92 Lang NP, Cumming BR, Löe H: Toothbrushing frequency as it relates to plaque development and gingival health. J Periodontol 1973;44:396–405.

93 Lindhe J, Nyman S: The effect of plaque control and surgical pocket elimination on the establishment and maintenance of periodontal health: a longitudinal study of periodontal therapy in cases of advanced disease. J Clin Periodontol 1975;2:67–79.

94 De la Rosa M, Zacarias Guerra J, Johnston DA, Radike AW: Plaque growth and removal with daily toothbrushing. J Periodontol 1979;50:661–664.

95 Apatzidou DA, Riggio MP, Kinane DF: Quadrant root planing versus same-day full-mouth root planing. II. Microbiological findings. J Clin Periodontol 2004;31:141–148.

96 Jervøe-Storm PM, AlAhdab H, Semaan E, Fimmers R, Jepsen S: Microbiological outcomes of quadrant versus full-mouth root planing as monitored by real-time PCR. J Clin Periodontol 2007;34:156–163.

97 De Soete M, Mongardini C, Peuwels M, Haffajee A, Socransky S, van Steenberghe D, Quirynen M: One-stage full-mouth disinfection: long-term microbiological results analyzed by checkerboard DNA-DNA hybridization. J Periodontol 2001;72:374–382.

98 Bollen CM, Vandekerckhove BN, Papaioannou W, Van Eldere J, Quirynen M: Full- versus partial-mouth disinfection in the treatment of periodontal infections. A pilot study: long-term microbiological observations. J Clin Periodontol 1996;23:960–970.

99 Quirynen M, Mongardini C, Pauwels M, Bollen CM, Van Eldere J, van Steenberghe D: One stage full- versus partial-mouth disinfection in the treatment of chronic adult or generalized early-onset periodontitis. II. Long-term impact on microbial load. J Periodontol 1999;70:646–656.

100 Zijnge V, Meijer HF, Lie MA, Tromp JA, Degener JE, Harmsen HJ, Abbas F: The recolonization hypothesis in a full-mouth or multiple-session treatment protocol: a blinded, randomized clinical trial. J Clin Periodontol 2010;37:518–525.

101 Lang NP, Tan WC, Krähenmann MA, Zwahlen M: A systematic review of the effects of full-mouth debridement with and without antiseptics in patients with chronic periodontitis. J Clin Periodontol 2008;35(suppl):8–21.

102 Braatz L, Garrett S, Claffey N, Egelberg J: Antimicrobial irrigation of deep pockets to supplement non-surgical periodontal therapy. II. Daily irrigation. J Clin Periodontol 1985;12:630–638.

103 MacAlpine R, Magnusson I, Kiger R, Crigger M, Garrett S, Egelberg J: Antimicrobial irrigation of deep pockets to supplement oral hygiene instruction and root debridement. I. Bi-weekly irrigation. J Clin Periodontol 1985;12:568–577.

104 Wennström JL, Heijl L, Dahlén G, Gröndahl K: Periodic subgingival antimicrobial irrigation of periodontal pockets. I. Clinical observations. J Clin Periodontol 1987;14:541–550.

105 Lander PE, Newcomb GM, Seymour GJ, Powell RN: The antimicrobial and clinical effects of a single subgingival irrigation of chlorhexidine in advanced periodontal lesions. J Clin Periodontol 1986;13:74–80.

106 Oosterwaal PJ, Mikx FH, van 't Hof MA, Renggli HH: Short-term bactericidal activity of chlorhexidine gel, stannous fluoride gel and amine fluoride gel tested in periodontal pockets. J Clin Periodontol 1991;18:97–100.

107 Sanz M, Teughels W, Group A of European Workshop on Periodontology. Innovations in non-surgical periodontal therapy: Consensus Report of the Sixth European Workshop on Periodontology. J Clin Periodontol 2008;35(suppl):3–7.

108 Herrera D, Alonso B, León R, Roldán S, Sanz M: Antimicrobial therapy in periodontitis: the use of systemic antimicrobials against the subgingival biofilm. J Clin Periodontol 2008;35(suppl):45–66.

109 Eberhard J, Jervøe-Storm PM, Needleman I, Worthington H, Jepsen S: Full-mouth treatment concepts for chronic periodontitis: a systematic review. J Clin Periodontol 2008;35:591–604.

Danae Anastasia Apatzidou, DDS, PhD
Department of Preventive Dentistry, Periodontology and Implant Biology, School of Dentistry
Aristotle University of Thessaloniki, University Campus
GR-54124 Thessaloniki (Greece)
Tel. +30 2310 278 844, E-Mail perioapatzidou@yahoo.gr

Kinane DF, Mombelli A (eds): Periodontal Disease. Front Oral Biol. Basel, Karger, 2012, vol 15, pp 117–132

Animal Models to Study Host-Bacteria Interactions Involved in Periodontitis

Dana T. Graves[a] · Jun Kang[a,b] · Oelisoa Andriankaja[a] · Keisuke Wada[a] · Carlos Rossa, Jr.[c]

[a]Department of Periodontics, School of Dental Medicine, University of Pennsylvania, Philadelphia, Pa., USA; [b]Department of Periodontology, Hospital and School of Stomatology, Peking University, Beijing, PR China; [c]Department of Diagnosis and Surgery, School of Dentistry at Araraquara, Universidade Estadual Pauliusta, Araraquara, Brazil

Abstract

Animal models have distinct advantages because they can mimic cellular complexities that occur in humans in vivo and are often more accurate than in vitro studies that take place on plastic surfaces with limited numbers of cell types present. Furthermore, cause and effect relationships can be established by applying inhibitors or activators or through the use of genetically modified animals. Such gain or loss of function studies are often difficult to achieve in human clinical studies, particularly in obtaining target tissue due to important ethical considerations. Animal models in periodontal disease are particularly important at this point in the development of the scientific basis for understanding the predominant pathological processes. Periodontal disease can be broken down into discrete steps, each of which may be studied separately depending upon the animal model. These steps involve the development of a pathogenic biofilm, invasion of connective tissue by bacteria or their products, induction of a destructive host response in connective tissue and limitation of a repair process that follows tissue breakdown. Animal studies can test hypotheses related to each of these steps, and should be evaluated by their capacity to test a specific hypothesis rather than recapitulating all aspects of periodontal disease. Thus, each of the models described below can be adapted to test discrete components of the pathological process of periodontal disease, but not necessarily all of them.

Animal models have limitations. However, the limitations are often less severe than those encountered during in vitro experiments, in which cells are typically examined on a plastic surface. Moreover, animal models often allow a more definitive analysis of cause and effect relationships than human clinical studies. A critical feature of animal models is the ability to examine complex host-bacteria interactions that cannot be carried out using single-cell populations under artificial conditions. This may be true even in the absence of a periodontium. For example, the impact of cytokines on the recruitment of inflammatory cells and the induction of bone resorption has been shown to occur in many different tissues, of which the periodontium is one. Thus, one can study the potential involvement of cytokines in the bone resorptive process in tissues ranging from the calvarium to the periodontium. This is possible since the functional role of particular mediators is typically consistent regardless of the specific bone examined. This raises the point that the value of animal studies should be considered in terms of their capacity to test specific hypotheses rather than their capacity to mimic all aspects of periodontal disease.

A problem in studying periodontal disease is the question of 'chronic' versus 'acute' nature of the inflammation. It is not known whether periodontal breakdown occurs in bursts or whether it is slow and continuous. At any point in time, the human periodontium adjacent to tooth has inflammatory cells recruited with characteristics of both chronic and acute inflammation. It is also clear that periodontal tissue breakdown involves both innate and acquired immunity [1–4]. The role of chronic and acute inflammation or innate versus acquired immunity can be studied using the appropriate animal model.

Animal models have characteristic features useful for investigating molecular mechanisms involved in the pathogenesis of periodontal diseases. Mouse models are particularly helpful because of the number of genetically modified strains that are available to study both gain of function and loss of function with regard to specific genes. Moreover, animal models provide an opportunity to investigate discrete steps of periodontal disease. Periodontal disease can be broken down into 5 different phases: (1) colonization; (2) invasion across the epithelium into connective tissue; (3) stimulation of an inflammatory response; (4) induction of connective tissue breakdown and bone resorption; (5) limitation of damage by repair processes, such as bone coupling, resulting in net bone loss. By selecting the appropriate animal model each of these steps can be investigated individually, whereas in human studies it is difficult to isolate a specific step and in vitro studies lack the complexity to examine specific phases. Moreover, a considerable amount is known regarding mouse and rat immune systems, and a wide range of reagents are available.

Rat Ligature Model

The induction of periodontal disease by the placement of ligatures around the teeth has been used in many different types of animals ranging from rats to nonhuman primates. Recently there has been considerable interest in the rat ligature model. Placement of a ligature leads to the accumulation of dental plaque and microulceration of the sulcular epithelium which, in turn, facilitate the invasion of periodontal pathogens into connective tissue. Loss of periodontal attachment and resorption of alveolar bone occurs predictably in a 7-day period in the rat [5–9], although some experiments have been conducted for much longer periods of time [10–13]. Similar to human periodontitis, bacterial penetration of the host tissue is thought to be a crucial factor affecting alveolar bone loss. The ligature alone does not initiate significant gingival inflammation or bone loss in gnotobiotic rats [12]. The role of bacteria in initiating periodontal destruction in the rat ligature model is supported by topical application of antiseptics, which reduce alveolar bone resorption, and systemic application of antibiotics, which reduces loss of attachment and bone resorption [11, 14, 15]. The role of bacteria in this model is further supported by findings that osteoclastogenesis and alveolar bone resorption are enhanced by the application of gram-negative bacteria [16].

As in human periodontitis, the stimulation of a host response induces periodontal destruction in the rat ligature model. This response involves the formation of gingival inflammation prior to bone resorption [17]. Periodontal destruction is decreased if the host response is reduced by inducing endotoxin tolerance [16]. The role of the inflammatory response is further demonstrated by reduced gingival inflammation, osteoclast formation and alveolar bone loss in the rat ligature model with application of prostaglandin inhibitors, including COX-2 specific inhibitors [7]. Low-dose doxycycline also reduces alveolar bone loss by inhibiting MMP activity [6]. In contrast, application of the cytokines interleukin (IL)-1 or TNF enhances bone loss in this model [18–22]. The ligature-induced model is also found to be sensitive to some systemic effects, such as smoking. Ligated rats exposed to cigarette smoke have greater bone loss and higher MMP-2 activity compared to animals with experimental periodontitis who were not exposed to cigarette smoke [23]. Similar results have been obtained when injecting nicotine into ligated rats [10, 24].

The use of the rat ligature model is useful in examining the effects of systemic disease on periodontal disease progression. For example, the impact of diabetes can be investigated by using type-2 Zucker diabetic fatty (ZDF) rats. ZDF rats develop diabetes by 8 weeks of age due to of obesity and insulin resistance [25]. With ligature-induced periodontitis, greater periodontal bone loss is found in the ZDF rats compared to matched normoglycemic controls. This is explained by the fact that diabetes enhances the duration and intensity of the inflammatory infiltrate, increasing osteoclastogenesis and thereby augmenting alveolar bone loss [26]. Furthermore, the effect of diabetes on bone coupling and its contribution to periodontal disease may be investigated after removing the ligatures. In this case, the amount of new alveolar bone formed following bone loss was 2.4- to 2.9-fold higher in normoglycemic rats compared with ZDF (diabetic) rats. The mechanisms for diminished bone coupling in diabetic animals may be explained by fewer bone lining and periodontal ligament cells, due in part to diabetes-enhanced apoptosis. Therefore, the rat ligature model is useful in investigating mechanisms by which a systemic condition such as diabetes affects the periodontium.

Rat *A. actinomycetemcomitans* Infection Model

An important aspect of periodontal disease is bacterial colonization of the tooth surface and its invasion into connective tissue. A problem with many rodent models of periodontal disease is that the bacteria used to induce the disease process only transiently infect the oral cavity since rodents are not natural hosts for many human bacteria. A well-documented exception to this general principle is infection of the rat with *Aggregatibacter actinomycetemcomitans*. Known natural hosts of *A. actinomycetemcomitans* include humans, old-world monkeys, cows and rats [27]. However, caution is still warranted since many laboratory strains of *A. actinomycetemcomitans* have a 'smooth' phenotype and greatly diminished capacity to adhere to surfaces, which is in sharp contrast to the *A. actinomycetemcomitans* found in fresh clinical isolates [28]. Therefore, the utility of bacterial laboratory strains are open to question, just as results from immortalized mammalian cells may need to be confirmed with primary cell cultures. This point is illustrated when the capacity of the smooth strain of *A. actinomycetemcomitans* to infect the oral cavity was found to be limited compared to recent clinical isolates [29]. Since *A. actinomycetemcomitans* naturally colonizes the oral cavity of rats, it can be examined to establish virulence factors critical for colonization [30, 31].

Wild-type *A. actinomycetemcomitans* adheres to rat buccal epithelial cells and is a common inhabitant of rice rats, while it is difficult to find naturally in Sprague Dawley rats from commercial sources [27]. However, *A. actinomycetemcomitans* can colonize the latter. On the other hand, *A. actinomycetemcomitans* has difficulty infecting the oral cavity of mice. Therefore, the mouse is not an adequate model to study *A. actinomycetemcomitans* colonization. Studies with *A. actinomycetemcomitans*-induced bone loss in mice are apparently due to transient infections that are self-limiting.

Several important factors should be considered when assessing colonization, including any reduction in the existing flora. The existing oral flora should be reduced to facilitate the addition of an exogenous bacterium. It is advantageous to tag *A. actinomycetemcomitans* as a means of distinguishing it from the existing flora, for example by using a rifampicin-resistant strain. To avoid transfer of microorganisms from one animal to another, they should be housed in separate cages. One method to consistently inoculate the oral cavities of rats with *A. actinomycetemcomitans* is to include known amounts in food, as described by Fine et al. [29].

The *A. actinomycetemcomitans* model has also been used to examine periodontal bone resorption and the systemic host response to infection. Li et al. [32] examined the role of B and CD4+ T cells in adaptive immunity of rats infected with *A. actinomycetemcomitans*. Results showed that elevated lymphocyte counts in rats inoculated with *A. actinomycetemcomitans* and the number of B and CD4+ T cells from regional lymph nodes were increased. After 2 weeks, *A-actinomycetemcomitans*-specific IgG and IgG2 responses were significantly enhanced. mRNA levels of several cytokines were found to be elevated in the lymph nodes including IL-2, IL-1, TNF, CD40 ligand, FasL, RANKL and osteoprotegerin.

The *A. actinomycetemcomitans* oral inoculation model in the rat has been used to examine the impact of rat strains on colonization and alveolar bone loss, in particular the fawn-hooded hypertensive (FHH), Dahl/salt sensitive and Brown Norway strains. Results showed significantly higher *A. actinomycetemcomitans* colonization in FHH rats than in the other strains. Interestingly, *A. actinomycetemcomitans*-associated bone loss was only observed among the FHH strain.

Oral Gavage Model of Experimental Periodontitis

The introduction of human strains of bacteria by oral gavage (inoculation of bacteria into the oral cavity) and subsequent impact on the periodontium has been studied in various rodent models [33, 34]. Various bacterial strains associated with periodontitis in humans have been used in this model, including *Porphyromonas gingivalis* [3, 35–37], *A. actinomycetemcomitans* [38], *Tannerella forsythia* [39] and *Treponema denticola* [40]. Typically, mice or rats are given a known number of bacteria in a viscous suspension (2% carboxymethylcellulose) administered orally. In many cases, the oral infection by human bacteria is transient. PCR has frequently been used for detection of bacteria, which has a limitation in that the bacteria detected are not necessarily living. Nevertheless, *T. denticola* has been detected in mice 71 days after inoculation [40] and *P. gingivalis* has been detected at up to 11 weeks [41]; 45% of rats exposed to *P. gingivalis* and 80% exposed to *T. denticola or T. forsythia* were found to harbor these bacteria after 4–6 weeks [42]. Reproducibility and predictability of infection is enhanced by prior treatment with antibiotics to reduce the endogenous oral flora and with repeated inoculations of the organism. Typical inoculations are scheduled every other day for 1 week with 10^9 CFU, although the number of inoculations may be increased in mouse strains that are less susceptible to periodontal disease. Significant bone loss can be measured histologically, by macroscopic analysis or by micro-computed tomography. Alveolar bone resorption is usually assessed around the maxillary molars since induction of bone loss in the lower molars is slower due to the thicker cortical alveolar bone and wider buccolingual dimensions. In many reports, mice are euthanized 6 weeks after the final inoculation [3, 43, 44], although recent reports have shown alveolar bone loss can be detected as early as 2 weeks after the final oral challenge with bacteria or 3 weeks after initiating infection [45]. It has been reported that inoculation of bacteria induces an immune response that may be comparable to that associated with periodontal disease in humans. Migration of leukocytes (mononuclear cells and neutrophils) into the gingival connective tissues, proliferation of the junctional epithelium, increased vascularization and serum antibody (IgG1 and IgG2) levels against the periodontal pathogens have been reported in these models [41]. Increased inflammatory cytokine levels have been reported following bacterial inoculation, including elevated levels of TNF-α, IL-12 and IFN-γ [46] and decreased anti-inflammatory cytokines such as IL-10 [40].

Both rats and mice are susceptible to alveolar bone loss induced in the oral gavage model. However, there are differences in the susceptibility to experimental periodontal disease among the various strains of mice. BALB/c, AKR/J, DBA/2J and C3H/HeN mice are more susceptible than C57/Bl6, A/J, 129/J, SJL/J and C3H/HeJ [36, 47]. These differences are related to genetic variances among the strains affecting components of the immune response, such as the point mutation on Toll-like receptor 4 (TLR4) in C3H/HeJ mice or differences in adaptive immunity [48]. Normal genetic variation (as well as genetically engineered alterations) in mice provides a potentially rich avenue to establish a cause-and-effect relationship between periodontal disease and aspects of the host response. Similarly, it would be interesting to take advantage of different characteristics of different rat strains to study particular pathogenic aspects of periodontal diseases, such as deficiencies in innate or adaptive immunity and the hyperinflammatory phenotype.

Infection of the oral cavity by topical administration of bacteria has also been carried out in rats. Many of these studies have examined the Sprague Dawley strain [41, 42, 45]. There is some evidence that genetic variances in laboratory rats influence the pathogenesis of periodontal disease and may be used to investigate the possible interrelationships between periodontal disease and the host response, such as differences in Th1 cytokine expression between Lewis and Fisher 344 strains [49].

One aspect that has been discussed in the oral infection model is the use of a single bacterial species versus the use of two or more microorganisms associated with periodontal disease. The complexity of bacterial stimulation is supported by findings that the persistence of *P. gingivalis* in the oral cavity of rats 4 weeks after initial challenge is significantly increased from 45 to 80–100% when this bacterium is co-infected with *T. forsythia* and *T. denticola*. Alveolar bone loss is significantly greater in animals challenged with a polymicrobial oral infection rather than monoinfection [42, 50].

It should be kept in mind that the inoculated bacterium is not necessarily responsible for periodontal bone loss since the introduction of exogenous bacteria may alter the host's bacterial flora. Unless one works with germ-free animals, specific pathogen-free laboratory rodents are not necessarily free of bacteria that can induce periodontal breakdown. This reflects the current understanding that there is a fine line between harmless and opportunistic commensal bacteria that can stimulate destruction of the periodontium. Moreover, there are complex bacterial interactions that are not always apparent. For example, exogenous infection with a single species of periodontal pathogen like *P. gingivalis* may modify the oral biofilm of the animal by interacting with endogenous bacterial species. The alteration of the endogenous flora may contribute to the disease process rather than the direct action of *P. gingivalis*.

As with all other experimental models of periodontal disease, variations of the oral gavage model need to be considered in regard to the purpose of the study. The oral gavage model has been used to establish proof of principle in vaccine development and to examine the relative impact of different antigens of pathogenic bacteria. For instance, immunization against RgpA and Kgp gingipain peptides or intact RgpA gingipain reduces *P. gingivalis*-induced bone resorption [51, 52]. The oral gavage model has been used to mechanistically link *P. gingivalis*-induced periodontitis to systemic conditions. For example, oral infection of hyperlipidemic mice with *P. gingivalis* mice accelerates atherosclerotic plaque formation [53, 54] and evidence suggests that this microorganism may play a role in aortic aneurysms through a mechanism involving smooth-muscle-cell-dependent hyperplasia [55]. However, the oral gavage model in rodents may be limited in studying several important aspects of human periodontal disease, such as the colonization process or bacteria-bacteria interactions.

Lipopolysaccharide Injection Model

Gram-negative bacteria are thought to be important periodontal pathogens. The lipopolysaccharide (LPS) component of the cell wall of these microorganisms is a significant inflammatory stimulus that triggers an innate immune response. Thus, the injection of LPS into the gingival tissues is a model for examining how the innate immune response to this bacterial component induces inflammation to stimulate osteoclastogenesis and bone loss. This model produces a histopathological aspect similar to the other models and to that observed in established periodontitis in humans, characterized by increased infiltration of leukocytes, higher levels of proinflammatory cytokines, collagen degradation and alveolar bone resorption. Typically, a defined amount of purified bacterial LPS suspended into small microvolumes (1–6 μl) is injected into the gingival tissues surrounding the posterior teeth of either mice or rats. The technique is somewhat sensitive and requires the use of microsyringes for small volumes and thin needles (28–33 G). When performing injections in mice, the use of optical magnification is suggested to improve visualization. The injection site frequently used is the palatal aspect of the upper first molars, but some studies have also performed the injections on the interdental papilla between lower first and second molars [56, 57]. These injections are performed under general anesthesia, usually 3 times a week for the experimental period. Alveolar bone loss has been observed as early as 7 days after the start of injections of LPS from various microorganisms, including *Escherichia coli* [58], *A. actinomycetemcomitans* [47] and *Salmonella typhimurium* [56], suggesting that the source of LPS may not significantly affect this particular outcome. Most studies have used experimental periods of 3–8 weeks for both mice [59, 60] and rats [61, 62].

The origin of the LPS used, both in terms of the microorganism from which it was obtained and of the purification process, is an important variable to be considered. Even though LPS from most bacteria are considered activators of TLR4, many periodontopathogenic bacteria – including *T. forsythia, Prevotella intermedia, Prevotella nigrescens, Fusobacterium nucleatum, A. actinomycetemcomitans, Veillonella parvula* and *P. gingivalis* – also activate TLR2 [63, 64]. In fact, the ability of *P. gingivalis* to induce alveolar bone loss is nearly abolished in mice lacking TLR2 [65]. However,

TLR4 may also play a prominent role in stimulating bone resorption as the C3H/HeJ mouse strain, which lacks functional TLR4, has significantly less alveolar bone resorption than wild-type mice upon injections of *A. actinomycetemcomitans* LPS [47]. A direct comparison between *A. actinomycetemcomitans* and *P. gingivalis* LPS suggests that both stimulate inflammation and bone loss, but the latter is a little bit less potent than *A. actinomycetemcomitans* LPS [57]. In a similar model, injections of *E. coli* and *P. gingivalis* LPS were compared by injecting them into the scalp to stimulate inflammation and calvarial bone resorption. In this model, *P. gingivalis* LPS and *E. coli* LPS were both relatively potent in stimulating inflammation and bone loss at the site of injection, but *E. coli* LPS was more potent in stimulating systemic inflammation resulting from endotoxemia [66].

Depending on the purpose of the study, the purification process for obtaining LPS is also an important concern, since possible contamination with proteins or DNA may trigger activation of other receptors and pathways leading to inflammation and alveolar bone loss. This concern may be particularly important for 'home made' preparations of LPS. The bacterial strain, purification methods and even culture conditions may influence the biochemistry of LPS. The antigenic moiety of LPS is contained in its lipid region, termed lipid A. Variations in the lipid A region can modulate the immune response elicited and the pathogenic potential of the LPS [67].

Among the advantages of various injection models is the greater experimental control over the pathogenic stimulus, which is directly delivered to the tissues (i.e. no 'colonization variable') in a titratable manner. Several mediators may be studied, such as specific TLR agonists to investigate proof of principle regarding the role of TLRs in inflammation and tissue destruction. Alternatively, live or heat-killed bacteria can be injected to investigate host-bacteria interactions in the periodontium. These mediate or induce tissue destruction by activating different cell types present in the host tissues, including fibroblasts, lymphocytes, macrophages, dendritic cells and osteoblasts. Events stimulated downstream include the production and release of an array of inflammatory mediators by these activated cells, including prostaglandins, cytokines and MMPs. Thus, injection models may be useful for investigating the activation of signaling pathways.

Calvarial Model

The calvarial model was first designed and developed to study the effect of cytokines on osteoclastogenesis [68]. It has been employed to investigate the effect of bacteria on bone resorption [69], bone resorbing activity of dendritic cells [70] and host-bacteria interactions in connective tissue in vivo [71, 72]. In this model, the stimulus is injected into the subcutaneous connective tissue over the calvarium. Histologically classic inflammatory events occur, including the rapid expression of proinflammatory cytokines within a few hours and the recruitment of polymorphonuclear leukocytes

(PMNs) within 24 h [73]. Bone resorption is induced within 3–5 days depending upon the size of the stimulus [74, 75]. If the stimulus is small, there is an inflammatory event which resolves with relatively little tissue destruction. A large inoculum creates a perturbation that induces a soft tissue wound and considerable bone resorption that subsequently undergoes repair. New bone formation can then be examined as part of the healing process and can be measured 8–12 days following inoculation of the stimulus [74, 76, 77]. The precise time frame in which events are measured may be modified by the strain of mouse [78]. The calvarial model is useful in examining the sequence of inflammation, destruction and repair. This may be critical to study periodontal disease where the net loss of tissue may be related to loss of balance between destruction of bone and connective tissue and the degree of repair that occurs.

An advantage of the calvarial model is that a sufficient amount of tissue is obtained for various profiling techniques. In one example, *P. gingivalis, T. denticola* and *T. forsythia* were injected into the connective tissue overlying calvarial bone in mice, and mRNA profiling was performed. This led to the identification of a number of inflammatory mediators that were induced in a bacteria-specific pattern [79–81].

The calvarial model is also well adapted to examine cause-and-effect relationships. In one example, the role of TNF-α in mediating the effects of *P. gingivalis* infection were examined by comparing TNF-receptor-ablated mice with wild-type controls. It was determined that formation of PMN infiltrate and osteoclastogenesis was dependent upon the induction of TNF receptor signaling. This suggests that the TNF activity plays an essential part in the inflammatory infiltrate formed in response to *P. gingivalis*. Although many bacterial components are chemotactic for PMNs, they are assumed to be less significant than bacteria-induced expression of cytokines, particularly TNF. In the same example, TNF activity also played an important role in the stimulation of fibroblast cell death stimulated by *P. gingivalis*. When *P. gingivalis* was inoculated in TNF-receptor-deficient mice, a 75% reduction in fibroblast apoptosis and a 60% decrease in osteoclast formation resulted compared to the same inoculation in wild-type mice [72]. A similar approach was used to establish the importance of TLR signaling pathways in response to LPS [77] and to demonstrate that dendritic cells have an important role in bone-resorption-associated inflammatory stimuli [70]. Likewise, the calvarial model has been used to investigate the functional importance of apoptosis in bacteria-induced bone resorption [76, 82]. When apoptosis was limited, there was a much more robust healing response to a large bacterial inoculum, particularly in diabetic animals. The anabolic effect of fluoxetine following bone resorption was demonstrated using the calvarial model [78]. Thus, after an episode of bone resorption, fluoxetine induced greater reparative bone formation, i.e. bone coupling, than that seen in vehicle-treated mice. This study also illustrates the necessity of using histological methods of analysis to better understand mechanisms. When examined by micro-computed tomography alone, it was not clear whether fluoxetine diminished bone resorption or enhanced coupled bone formation. According to the histological analysis, the more dominant effect was upon the coupled bone formation.

The other benefit of using the calvarial model is the capability to deliver a precise amount of inoculum and examine qualitative changes in the response to different levels of the same stimulus. This has been studied in the calvarial model using mice with targeted deletions as well as inhibitory antibodies. It has been indicated that osteoclastogenesis is induced by high doses of enteric LPS and substantially mediated by IL-1 and TNF receptor signaling. On the other hand, lower doses of LPS induce osteoclastogenesis through other pathways that are independent of TNF and IL-1 and involve IL-11 [83, 84].

In addition to studying bone repair, soft tissue healing can also be investigated in the calvarial model. The microarray studies mentioned above indicated that the gene transcription profiles in inflamed calvarial bone induced by microbial infection were different from the profiles observed in the soft tissue [79–81]. Another study has demonstrated the use of this model to examine a bacterially induced soft tissue wound that follows the classic steps of healing, including the migration of fibroblasts into the wounded site, proliferation, production of extracellular matrix and reorganization [85]. The repair of bacteria-associated wounds in diabetic mice is related to the greater production of TNF. By use of the calvarial model, it was shown that a TNF-specific inhibitor considerably enhanced the repair process following injection of *P. gingivalis* [76]. Indeed, there is a significant improvement in soft tissue healing when fibroblast apoptosis is reduced with a caspase inhibitor [82]. This may yield further insights into the periodontium, where the loss of fibroblasts is one of the most distinctive cellular changes that occurs in progressive periodontal disease [86]. Thus, studies in the calvarial model point to potential novel differences in the response to a bacterially induced injury in diabetic animals.

As mentioned above, the capacity to control an infection by inoculation allows precise examination of the induced host response. This can be further enlarged to study the effect with or without the presence of the adaptive immune response. Mice are naïve to *P. gingivalis*. By immunizing mice before injection of bacteria into the calvarial connective tissue, the contribution of the adaptive immune response can be examined by comparison with non-immunized mice [71]. A striking feature following *P. gingivalis* inoculation was the enhanced expression of bone-resorptive cytokines, osteoclastogenesis and bone loss in the immunized group, suggesting that the adaptive immune response contributes to the destruction of bone stimulated by *P. gingivalis*. Moreover, reparative coupled bone formation was significantly inhibited by the addition of the adaptive immune response [87].

Rat Mandibular Critical Size Defect

Knowledge of bone formation is important in understanding periodontal disease and treatment. It has been proposed that a failure to form bone after an episode of periodontal bone resorption (from uncoupling) is an important component of periodontitis [88]. Therefore, models of bone formation are important in discussing the spectrum of various periodontal animal models. The rat mandibular bone defect model was developed to study bone formation in the mandible [89]. This model has been useful in designing studies to compare the efficacy of different materials to augment bone formation including barrier membranes [90–92], bone grafts [93], growth factors and hormones [92, 94–96] and other factors such as ultrasound stimulation [97, 98]. In this model, the angle of the rat mandible is surgically exposed on both the buccal and the lingual aspects and a through-and-through osseous defect of 5 mm is created with a trephine burr. A similar defect may be created on the contralateral side, which may be used as a control if specific bone augmentation material is being tested.

In recent years, micro-computed tomography has been used as a fast precise non-destructive analytical procedure to measure bone volume within the defect created. There is a significant correlation ($p < 0.0001$) between micro-computed tomography values and histomorphometric analysis in measuring the amount of new bone and grafted material after 3 weeks of early healing [99, 100]. The rat mandibular bone formation model has been used to compare the performance of a degradable synthetic barrier membrane PDLLCL [poly(D-,L-lactide-ε-caprolactone)] with a comparable type 1 collagen or expanded ePTFE (polytetrafluoroethylene) membrane [101]. It was found that significantly more bone formation was observed under type 1 collagen and ePTFE membranes than the PDLLCL membranes. This model has been used to compare bone grafting materials β-tricalcium phosphate (β-TCP), allogenic bone and allogenic bone combined with β-TCP [93]. In this study, the authors concluded that β-TCP absorption occurred at a slower rate and enhanced bone formation when combined with an allograft because it helped maintain space more effectively than the allograft alone.

One of the advantages of the mandibular-angle-defect model is the ability to investigate the efficacy of locally and systemically delivered biologics. Local delivery of growth factors has been investigated in the rat mandibular defect model to compare bone morphogenetic protein-2 (BMP-2) with and without barrier membrane [94]. Interestingly, the results indicated that BMP-2 without membrane had significantly more bone repair than BMP-2 covered with membrane. It was suggested that the barrier membrane blocked a source of stem cells from migrating into the defect and limiting the amount of new bone formation. Since then, other studies with a growth factor have shown similar tendencies, which has led to recommendations that biologics which enhance bone formation should not be used with barrier membranes. This model has also been used to evaluate the effectiveness of systemically delivered calcitonin in promoting bone growth in surgical bone defects in rat mandibles [96]. The results revealed the calcitonin reduced bone formation at 1, 2 and 3 weeks of healing compared to the non-calcitonin-treated group. However, no significant difference was seen after 4 weeks of healing. Thus, bone repair in the calcitonin-treated animals appeared to be delayed in comparison to controls. Thus, the rat mandibular critical size defect is useful and effective in assessing factors and materials that promote bone formation.

References

1 Assuma R, Oates T, Cochran D, Amar S, Graves D: IL-1 and TNF antagonists inhibit the inflammatory response and bone loss in experimental periodontitis. J Immun 1998;160:403–409.

2 Teng Y, Nguyen H, Gao X, Kong Y, Gorczynski R, Singh B, Ellen R, Penninger J: Functional human T-cell immunity and osteoprotegerin ligand control alveolar bone destruction in periodontal infection. J Clin Invest 2000;106:R59–R67.

3 Baker P, Dixon M, Evans R, Dufour L, Johnson E, Roopenian D: CD4(+) T cells and the proinflammatory cytokines gamma interferon and interleukin-6 contribute to alveolar bone loss in mice. Infect Immun 1999;67:2804–2809.

4 Delima A, Spyros K, Amar S, Graves DT: Inflammation and tissue loss caused by periodontal pathogens is reduced by IL-1 antagonists. J Infect Dis 2002;186:511–516.

5 Nowotny A, Sanavi F: Induction of nonspecific tolerance to endotoxins reduces the alveolar bone resorption in ligature-treated rats. Infect Immun1983;39:873–878.

6 Bezerra MM, Brito GA, Ribeiro RA, Rocha FA: Low-dose doxycycline prevents inflammatory bone resorption in rats. Braz J Med Biol Res 2002;35:613–616.

7 Bezerra MM, de Lima V, Alencar VB, Vieira IB, Brito GA, Ribeiro RA, Rocha FA: Selective cyclooxygenase-2 inhibition prevents alveolar bone loss in experimental periodontitis in rats. J Periodontol 2000;71:1009–1014.

8 Lohinai Z, Benedek P, Feher E, Gyorfi A, Rosivall L, Fazekas A, Salzman AL, Szabo C: Protective effects of mercaptoethylguanidine, a selective inhibitor of inducible nitric oxide synthase, in ligature-induced periodontitis in the rat. Br J Pharmacol 1998;123: 353–360.

9 Xie R, Kuijpers-Jagtman AM, Maltha JC: Inflammatory responses in two commonly used rat models for experimental tooth movement: comparison with ligature-induced periodontitis. Arch Oral Biol 2011;56:159–167.

10 Nociti FH Jr, Nogueira-Filho GR, Tramontina VA, Machado MA, Barros SP, Sallum EA, Sallum AW: Histometric evaluation of the effect of nicotine administration on periodontal breakdown: an in vivo study. J Periodontal Res 2001;36:361–366.

11 Weiner GS, DeMarco TJ, Bissada NF: Long term effect of systemic tetracycline administration on the severity of induced periodontitis in the rat. J Periodontol 1979;50:619–623.

12 Rovin S, Costich ER, Gordon HA: The influence of bacteria and irritation in the initiation of periodontal disease in germfree and conventional rats. J Periodontal Res 1966;1:193–204.

13 Kuhr A, Popa-Wagner A, Schmoll H, Schwahn C, Kocher T: Observations on experimental marginal periodontitis in rats. J Periodontal Res 2004;39:101–106.

14 Kenworthy R, Baverel M: Studies of a periodontal tissue lesion in the rat, untreated or treated with chlorhexidine digluconate. J Clin Periodontol 1981; 8:349–358.

15 Luan Q, Desta T, Chehab L, Sanders VJ, Plattner J, Graves DT: Inhibition of experimental periodontitis by a topical boron-based antimicrobial. J Dent Res 2008;87:148–152.

16 Samejima Y, Ebisu S, Okada H: Effect of infection with *Eikenella corrodens* on the progression of ligature-induced periodontitis in rats. J Periodontal Res 1990;25:308–315.

17 Gyorfi A, Fazekas A, Suba Z, Ender F, Rosivall L: Neurogenic component in ligature-induced periodontitis in the rat. J Clin Periodontol 1994;21:601–605.

18 Koide M, Suda S, Saitoh S, Ofuji Y, Suzuki T, Yoshie H, Takai M, Ono Y, Taniguchi Y, Hara K: In vivo administration of IL-1 beta accelerates silk ligature-induced alveolar bone resorption in rats. J Oral Pathol Med 1995;24:420–434.

19 Gaspersic R, Stiblar-Martincic D, Osredkar J, Skaleric U: Influence of subcutaneous administration of recombinant TNF-alpha on ligature-induced periodontitis in rats. J Periodontal Res 2003;38:198–203.

20 Tomofuji T, Ekuni D, Irie K, Azuma T, Endo Y, Tamaki N, Sanbe T, Murakami J, Yamamoto T, Morita M: Preventive effects of a cocoa-enriched diet on gingival oxidative stress in experimental periodontitis. J Periodontol 2009;80:1799–1808.

21 Ekuni D, Tomofuji T, Irie K, Kasuyama K, Umakoshi M, Azuma T, Tamaki N, Sanbe T, Endo Y, Yamamoto T, Nishida T, Morita M: Effects of periodontitis on aortic insulin resistance in an obese rat model. Lab Invest 2010;90:348–359.

22 Endo Y, Tomofuji T, Ekuni D, Irie K, Azuma T, Tamaki N, Yamamoto T, Morita M: Experimental periodontitis induces gene expression of proinflammatory cytokines in liver and white adipose tissues in obesity. J Periodontol 2010;81:520–526.

23 Cesar Neto J, de Souza A, Barbieri D, Moreno HJ, Sallum EA, Nociti FH Jr: Matrix metalloproteinase-2 may be involved with increased bone loss associated with experimental periodontitis and smoking: a study in rats. J Periodontol 2004;75:995–1000.

24 Benatti BB, Nogueira-Filho GR, Diniz MC, Sallum EA, Sallum AW, Nociti FH Jr: Stress may enhance nicotine effects on periodontal tissues: an in vivo study in rats. J Periodontal Res 2003;38:351–353.

25 Clark J, Palmer C, Shaw W: The diabetic Zucker fatty rat. Proc Soc Exp Biol Med 1983;173:68–75.

26 Liu R, Bal HS, Desta T, Krothapalli N, Alyassi M, Luan Q, Graves D: Diabetes enhances periodontal bone loss through enhanced resorption and diminished bone formation. J Dental Res 2006;85:510–514.

27 Fine DH, Velliyagounder K, Furgang D, Kaplan JB: The *Actinobacillus actinomycetemcomitans* autotransporter adhesin Aae exhibits specificity for buccal epithelial cells from humans and old world primates. Infect Immun 2005;73:1947–1953.

28 Scannapieco FA, Millar SJ, Reynolds HS, Zambon JJ, Levine MJ: Effect of anaerobiosis on the surface ultrastructure and surface proteins of *Actinobacillus actinomycetemcomitans (Haemophilus actinomycetemcomitans)*. Infect Immun 1987;55:2320–2323.

29 Fine DH, Goncharoff P, Schreiner H, Chang KM, Furgang D, Figurski D: Colonization and persistence of rough and smooth colony variants of *Actinobacillus actinomycetemcomitans* in the mouths of rats. Arch Oral Biol 2001;46:1065–1078.

30 Kachlany SC, Planet PJ, DeSalle R, Fine DH, Figurski DH: Genes for tight adherence of *Actinobacillus actinomycetemcomitans*: from plaque to plague to pond scum. Trends Microbiol 2001;9: 429–437.

31 Saito T, Ishihara K, Ryu M, Okuda K, Sakurai K: Fimbriae-associated genes are biofilm-forming factors in *Aggregatibacter actinomycetemcomitans* strains. Bull Tokyo Dent Coll 2010;51:145–150.

32 Li Y, Messina C, Bendaoud M, Fine DH, Schreiner H, Tsiagbe VK: Adaptive immune response in osteoclastic bone resorption induced by orally administered *Aggregatibacter actinomycetemcomitans* in a rat model of periodontal disease. Mol Oral Microbiol 2010;25:275–292.

33 Klausen B, Evans RT, Ramamurthy NS, Golub LM, Sfintescu C, Lee JY, Bedi G, Zambon JJ, Genco RJ: Periodontal bone level and gingival proteinase activity in gnotobiotic rats immunized with *Bacteroides gingivalis*. Oral Microbiol Immunol 1991;6:193–201.

34 Chang KM, Ramamurthy NS, McNamara TF, Genco RJ, Golub LM: Infection with a gram-negative organism stimulates gingival collagenase production in non-diabetic and diabetic germfree rats. J Periodontal Res 1988;23:239–244.

35 Baker P, Evans R, Roopenian D: Oral infection with *Porphyromonas gingivalis* and induced alveolar bone loss in immunocompetent and severe combined immunodeficient mice. Arch Oral Biol 1994; 39:1035–1040.

36 Baker PJ, Dixon M, Roopenian DC: Genetic control of susceptibility to Porphyromonas gingivalis-induced alveolar bone loss in mice. Infect Immun 2000;68:5864–5868.

37 Lalla E, Lamster I, Feit M, Huang L, Schmidt A: A murine model of accelerated periodontal disease in diabetes. J Periodontal Res 1998;33:387–399.

38 Garlet GP, Cardoso CR, Silva TA, Ferreira BR, Avila-Campos MJ, Cunha FQ, Silva JS: Cytokine pattern determines the progression of experimental periodontal disease induced by *Actinobacillus actinomycetemcomitans* through the modulation of MMPs, RANKL, and their physiological inhibitors. Oral Microbiol Immunol 2006;21:12–20.

39 Sharma A, Inagaki S, Honma K, Sfintescu C, Baker PJ, Evans RT: *Tannerella forsythia*-induced alveolar bone loss in mice involves leucine-rich-repeat BspA protein. J Dent Res 2005;84:462–467.

40 Lee SF, Andrian E, Rowland E, Marquez IC: Immune response and alveolar bone resorption in a mouse model of *Treponema denticola* infection. Infect Immun 2009;77:694–698.

41 Bainbridge B, Verma RK, Eastman C, Yehia B, Rivera M, Moffatt C, Bhattacharyya I, Lamont RJ, Kesavalu L: Role of *Porphyromonas gingivalis* phosphoserine phosphatase enzyme SerB in inflammation, immune response, and induction of alveolar bone resorption in rats. Infect Immun 2010;78:4560–4569.

42 Kesavalu L, Sathishkumar S, Bakthavatchalu V, Matthews C, Dawson D, Steffen M, Ebersole JL: Rat model of polymicrobial infection, immunity, and alveolar bone resorption in periodontal disease. Infect Immun 2007;75:1704–1712.

43 Baker PJ, Dixon M, Evans RT, Roopenian DC: Heterogeneity of *Porphyromonas gingivalis* strains in the induction of alveolar bone loss in mice. Oral Microbiol Immunol 2000;15:27–32.

44 Yu JJ, Ruddy MJ, Wong GC, Sfintescu C, Baker PJ, Smith JB, Evans RT, Gaffen SL: An essential role for IL-17 in preventing pathogen-initiated bone destruction: recruitment of neutrophils to inflamed bone requires IL-17 receptor-dependent signals. Blood 2007;109:3794–3802.
45 Okada Y, Hamada N, Kim Y, Takahashi Y, Sasaguri K, Ozono S, Sato S: Blockade of sympathetic b-receptors inhibits *Porphyromonas gingivalis*-induced alveolar bone loss in an experimental rat periodontitis model. Arch Oral Biol 2010;55:502–508.
46 Garlet GP, Avila-Campos MJ, Milanezi CM, Ferreira BR, Silva JS: *Actinobacillus actinomycetemcomitans*-induced periodontal disease in mice: patterns of cytokine, chemokine, and chemokine receptor expression and leukocyte migration. Microbes Infect 2005;7:738–747.
47 Nakamura H, Fukusaki Y, Yoshimura A, Shiraishi C, Kishimoto M, Kaneko T, Hara Y: Lack of Toll-like receptor 4 decreases lipopolysaccharide-induced bone resorption in C3H/HeJ mice in vivo. Oral Microbiol Immunol 2008;23:190–195.
48 Gemmell E, Winning TA, Carter CL, Ford PJ, Bird PS, Ashman RB, Grieco DA, Seymour GJ: Differences in mouse strain influence leukocyte and immunoglobulin phenotype response to *Porphyromonas gingivalis*. Oral Microbiol Immunol 2003;18:364–370.
49 Breivik T, Thrane PS, Gjermo P, Opstad PK, Pabst R, von Horsten S: Hypothalamic- pituitary-adrenal axis activation by experimental periodontal disease in rats. J Periodontal Res 2001;36:295–300.
50 Polak J, Hench L: Gene therapy progress and prospects: in tissue engineering. Gene Ther 2005;12: 1725–1733.
51 O'Brien-Simpson NM, Pathirana RD, Paolini RA, Chen YY, Veith PD, Tam V, Ally N, Pike RN, Reynolds EC: An immune response directed to proteinase and adhesin functional epitopes protects against *Porphyromonas gingivalis*-induced periodontal bone loss. J Immunol 2005;175:3980–3989.
52 Gibson FC 3rd, Genco CA: Prevention of *Porphyromonas gingivalis*-induced oral bone loss following immunization with gingipain R1. Infect Immun 2001;69:7959–7963.
53 Lalla E, Lamster IB, Hofmann MA, Bucciarelli L, Jerud AP, Tucker S, Lu Y, Papapanou PN, Schmidt AM: Oral infection with a periodontal pathogen accelerates early atherosclerosis in apolipoprotein E-null mice. Arterioscler Thromb Vasc Biol 2003;23: 1405–1411.
54 Gibson FC 3rd, Hong C, Chou HH, Yumoto H, Chen J, Lien E, Wong J, Genco CA: Innate immune recognition of invasive bacteria accelerates atherosclerosis in apolipoprotein E-deficient mice. Circulation 2004;109:2801–2806.
55 Wada K, Kamisaki Y: Roles of oral bacteria in cardiovascular diseases – from molecular mechanisms to clinical cases: involvement of *Porphyromonas gingivalis* in the development of human aortic aneurysm. J Pharmacol Sci 2010;113:115–119.
56 Dumitrescu AL, Abd-El-Aleem S, Morales-Aza B, Donaldson LF: A model of periodontitis in the rat: effect of lipopolysaccharide on bone resorption, osteoclast activity, and local peptidergic innervation. J Clin Periodontol 2004;31:596–603.
57 Nishida E, Hara Y, Kaneko T, Ikeda Y, Ukai T, Kato I: Bone resorption and local interleukin-1alpha and interleukin-1beta synthesis induced by *Actinobacillus actinomycetemcomitans* and *Porphyromonas gingivalis* lipopolysaccharide. J Periodontal Res 2001;36:1–8.
58 Buduneli E, Vardar-Sengul S, Buduneli N, Atilla G, Wahlgren J, Sorsa T: Matrix metalloproteinases, tissue inhibitor of matrix metalloproteinase-1, and laminin-5 gamma2 chain immunolocalization in gingival tissue of endotoxin-induced periodontitis in rats: effects of low-dose doxycycline and alendronate. J Periodontol 2007;78:127–134.
59 Sartori R, Li F, Kirkwood KL: MAP kinase phosphatase-1 protects against inflammatory bone loss. J Dent Res 2009;88:1125–1130.
60 Nakamura H, Ukai T, Yoshimura A, Kozuka Y, Yoshioka H, Yoshinaga Y, Abe Y, Hara Y: Green tea catechin inhibits lipopolysaccharide-induced bone resorption in vivo. J Periodontal Res 2010;45:23–30.
61 Garcia de Aquino S, Manzolli Leite FR, Stach-Machado DR, Francisco da Silva JA, Spolidorio LC, Rossa C Jr: Signaling pathways associated with the expression of inflammatory mediators activated during the course of two models of experimental periodontitis. Life Sci 2009;84:745–754.
62 Rogers JE, Li F, Coatney DD, Rossa C, Bronson P, Krieder JM, Giannobile WV, Kirkwood KL: *Actinobacillus actinomycetemcomitans* lipopolysaccharide-mediated experimental bone loss model for aggressive periodontitis. J Periodontol 2007;78: 550–558.
63 Darveau RP, Pham TT, Lemley K, Reife RA, Bainbridge BW, Coats SR, Howald WN, Way SS, Hajjar AM: *Porphyromonas gingivalis* lipopolysaccharide contains multiple lipid A species that functionally interact with both toll-like receptors 2 and 4. Infect Immun 2004;72:5041–5051.

64 Kikkert R, Laine ML, Aarden LA, van Winkelhoff AJ: Activation of toll-like receptors 2 and 4 by gram-negative periodontal bacteria. Oral Microbiol Immunol 2007;22:145–151.
65 Burns E, Bachrach G, Shapira L, Nussbaum G: Cutting Edge: TLR2 is required for the innate response to *Porphyromonas gingivalis*: activation leads to bacterial persistence and TLR2 deficiency attenuates induced alveolar bone resorption. J Immunol 2006;177:8296–8300.
66 Liu R, Desta T, Raptis M, Darveau RP, Graves DT: *P. gingivalis* and *E. coli* lipopolysaccharides exhibit different systemic but similar local induction of inflammatory markers. J Periodontol 2008;79:1241–1247.
67 Berezow AB, Ernst RK, Coats SR, Braham PH, Karimi-Naser LM, Darveau RP: The structurally similar, penta-acylated lipopolysaccharides of *Porphyromonas gingivalis* and *Bacteroides* elicit strikingly different innate immune responses. Microb Pathog 2009;47:68–77.
68 Boyce B, Aufdemorte T, Garrett I, Yates A, Mundy G: Effects of interleukin-1 on bone turnover in normal mice. Endocrinology 1989;125:1142–1150.
69 Zubery Y, Dunstan C, Story B, Kesavalu L, Ebersole J, Holt S, Boyce B: Bone resorption caused by three periodontal pathogens in vivo in mice is mediated in part by prostaglandin. Infect Immun 1998;66:4158–4162.
70 Maitra R, Follenzi A, Yaghoobian A, Montagna C, Merlin S, Cannizzo ES, Hardin JA, Cobelli N, Stanley ER, Santambrogio L: Dendritic cell-mediated in vivo bone resorption. J Immunol 2010;185:1485–1491.
71 Leone CW, Bokhadhoor H, Kuo D, Desta T, Yang J, Siqueira MF, Amar S, Graves DT: Immunization enhances inflammation and tissue destruction in response to *Porphyromonas gingivalis*. Infect Immun 2006;74:2286–2292.
72 Graves D, Oskoui M, Volejnikova S, Naguib G, Cai S, Desta T, Kakouras A, Jiang Y: Tumor necrosis factor modulates fibroblast apoptosis, PMN recruitment, and osteoclast formation in response to *P. gingivalis* infection. J Dent Res 2001;80:1875–1879.
73 Graves D, Naguib G, Lu H, Desta T, Amar S: *Porphyromonas gingivalis* fimbriae are proinflammatory but do not play a prominent role in the innate immune response to *P. gingivalis*. J Endotoxin Res 2005;11:13–18.
74 He H, Liu R, Desta T, Leone C, Gerstenfeld L, Graves D: Diabetes causes decreased osteoclastogenesis, reduced bone formation, and enhanced apoptosis of osteoblastic cells in bacteria stimulated bone loss. Endocrinology 2004;145:447–452.
75 Li L, Khansari A, Shapira L, Graves D, Amar S: Contribution of interleukin-11 and protaglandin(s) in lipopolysaccharide-induced bone resorption in vivo. Infect Immun 2002;70:3915–3922.
76 Liu R, Bal HS, Desta T, Behl Y, Graves DT: Tumor necrosis factor-alpha mediates diabetes-enhanced apoptosis of matrix-producing cells and impairs diabetic healing. Am J Pathol 2006;168:757–764.
77 Nason R, Lee DH, Jung JY, Chole RA: Radiographic and micro-computed tomographic imaging of lipopolysaccharide-mediated bone resorption. Ann Otol Rhinol Laryngol 2009;118:391–396.
78 Battaglino R, Vokes M, Schulze-Spate U, Sharma A, Graves D, Kohler T, Muller R, Yoganathan S, Stashenko P: Fluoxetine treatment increases trabecular bone formation in mice. J Cell Biochem 2007;100:1387–1394.
79 Bakthavatchalu V, Meka A, Sathishkumar S, Lopez MC, Verma RK, Wallet SM, Bhattacharyya I, Boyce BF, Mans JJ, Lamont RJ, Baker HV, Ebersole JL, Kesavalu L: Molecular characterization of *Treponema denticola* infection-induced bone and soft tissue transcriptional profiles. Mol Oral Microbiol 2010;25:260–274.
80 Meka A, Bakthavatchalu V, Sathishkumar S, Lopez MC, Verma RK, Wallet SM, Bhattacharyya I, Boyce BF, Handfield M, Lamont RJ, Baker HV, Ebersole JL, Kesavalu L: *Porphyromonas gingivalis* infection-induced tissue and bone transcriptional profiles. Mol Oral Microbiol 2010;25:61–74.
81 Bakthavatchalu V, Meka A, Sathishkumar S, Lopez MC, Bhattacharyya I, Boyce BF, Mans JJ, Lamont RJ, Baker HV, Ebersole JL, Kesavalu L: *Tannerella forsythia* infection-induced calvarial bone and soft tissue transcriptional profiles. Mol Oral Microbiol 2010;25:317–330.
82 Al-Mashat HA, Kandru S, Liu R, Behl Y, Desta T, Graves DT: Diabetes enhances mRNA levels of proapoptotic genes and caspase activity, which contribute to impaired healing. Diabetes 2006;55:487–495.
83 Chiang C, Kyritsis G, Graves D, Amar S: Interleukin-1 and tumor necrosis factor activities partially account for calvarial bone resorption induced by local injection of lipopolysaccharide. Infect Immun 1999;67:4231–4236.
84 Li H, Suzuki J, Bayna E, Zhang F, Dalle M, Clark A, Engler R, Lew W: Lipopolysaccharide induces apoptosis in adult rat ventricular myocytes via cardiac AT(1) receptors. Am J Physiol Heart Circ Physiol 2002;282:H461–H467.
85 Liu R, Desta T, He H, Graves D: Diabetes alters the response to bacteria by enhancing fibroblast apoptosis. Endocrinology 2004;145:2997–3003.

86 Zappa U, Reinking-Zappa M, Graf H, Chase D: Cell populations associated with active probing attachment loss. J Periodontol 1992;63:748–752.
87 Behl Y, Siqueira M, Ortiz J, Li J, Desta T, Faibish D, Graves DT: Activation of the acquired immune response reduces coupled bone formation in response to a periodontal pathogen. J Immunol 2008;181:8711–8718.
88 Graves DT, Li J, Cochran DL: Inflammation and uncoupling as mechanisms of periodontal bone loss. J Dent Res 2011;90:143–153.
89 Dahlin C, Linde A, Gottlow J, Nyman S: Healing of bone defects by guided tissue regeneration. Plast Reconstr Surg 1988;81:672–676.
90 Sandberg E, Dahlin C, Linde A: Bone regeneration by the osteopromotion technique using bioabsorbable membranes: an experimental study in rats. J Oral Maxillofac Surg 1993;51:1106–1114.
91 Bartee BK, Carr JA: Evaluation of a high-density polytetrafluoroethylene (n-PTFE) membrane as a barrier material to facilitate guided bone regeneration in the rat mandible. J Oral Implantol 1995;21: 88–95.
92 Deppe H, Stemberger A, Hillemanns M: Effects of osteopromotive and anti-infective membranes on bone regeneration: an experimental study in rat mandibular defects. Int J Oral Maxillofac Implants 2003;18:369–376.
93 Hirota M, Matsui Y, Mizuki N, Kishi T, Watanuki K, Ozawa T, Fukui T, Shoji S, Adachi M, Monden Y, Iwai T, Tohnai I: Combination with allogenic bone reduces early absorption of beta-tricalcium phosphate (beta-TCP) and enhances the role as a bone regeneration scaffold: experimental animal study in rat mandibular bone defects. Dent Mater J 2009;28: 153–161.
94 Hedner E, Linde A: Efficacy of bone morphogenetic protein (BMP) with osteopromotive membranes – an experimental study in rat mandibular defects. Eur J Oral Sci 1995;103:236–241.
95 Almeida JD, Arisawa EA, Balducci I, da Rocha RF, Carvalho YR: Homeopathic treatment for bone regeneration: experimental study. Homeopathy 2009;98:92–96.
96 Almeida JD, Arisawa EA, da Rocha RF, Carvalho YR: Effect of calcitonin on bone regeneration in male rats: a histomorphometric analysis. Int J Oral Maxillofac Surg 2007;36:435–440.
97 Schortinghuis J, Stegenga B, Raghoebar GM, de Bont LG: Ultrasound stimulation of maxillofacial bone healing. Crit Rev Oral Biol Med 2003;14:63–74.
98 Schortinghuis J, Ruben JL, Raghoebar GM, Stegenga B: Therapeutic ultrasound to stimulate osteoconduction: a placebo controlled single blind study using e-PTFE membranes in rats. Arch Oral Biol 2004;49:413–420.
99 Lu M, Rabie AB: Quantitative assessment of early healing of intramembranous and endochondral autogenous bone grafts using micro-computed tomography and Q-win image analyzer. Int J Oral Maxillofac Surg 2004;33:369–376.
100 Gielkens PF, Schortinghuis J, de Jong JR, Huysmans MC, Leeuwen MB, Raghoebar GM, Bos RR, Stegenga B: A comparison of micro-CT, microradiography and histomorphometry in bone research. Arch Oral Biol 2008;53:558–566.
101 Gielkens PF, Schortinghuis J, de Jong JR, Raghoebar GM, Stegenga B, Bos RR: Vivosorb, Bio-Gide, and Gore-Tex as barrier membranes in rat mandibular defects: an evaluation by microradiography and micro-CT. Clin Oral Implants Res 2008;19:516–521.

Dr. Dana T. Graves
Department of Periodontics, School of Dental Medicine, University of Pennsylvania
240 S. 40th Street
Philadelphia, PA 19104 (USA)
Tel. +1 215 898 9068, E-Mail dtgraves@dental.upenn.edu

Kinane DF, Mombelli A (eds): Periodontal Disease. Front Oral Biol. Basel, Karger, 2012, vol 15, pp 133–148

Antimicrobial Advances in Treating Periodontal Diseases

Andrea Mombelli

Department of Periodontology and Oral Pathophysiology, School of Dental Medicine, University of Geneva, Geneva, Switzerland

Abstract

Antibiotics are generally an efficient means of treating bacterial infections, and therefore are an obvious candidate in the treatment of periodontal diseases. Systemically and locally administered antimicrobial agents of all kinds have been evaluated in multiple clinical trials. The vast majority of studies have tested antibiotics as adjuncts to non-surgical debridement. No regime has demonstrated superiority over systemically administered amoxicillin and metronidazole in the treatment of any clinically or microbiologically defined variant of periodontal disease. The frequency and consequences of adverse effects of antibiotics have always to be balanced against the potential consequences of not rapidly suppressing a periodontal infection. Proposed strategies to reduce the risk of bacterial antimicrobial resistance include: prescribing two drugs with a synergistic or complementary effect, the administration of antibiotics at a high dose for a short period, a combined approach with mechanical debridement to disrupt biofilms, and the focus on therapeutic rather than prophylactic use. Derivatives of existing antibiotic classes and new compounds that act on unique targets are the subject of preclinical investigations with a focus on action against antibiotic-resistant medical pathogens. In light of the excellent results of a combination therapy with well-established drugs that are cheap and efficient, clinical trials should compare newly proposed protocols for periodontal therapy to a positive control. Future studies should focus not only on the action against the microorganisms directly involved in periodontal diseases, but also on those relevant to other medical concerns.

Periodontal diseases are commonly viewed as the negative consequence of the interaction between microorganisms colonizing tooth surfaces and the soft tissues anchoring the teeth in the jaw. Treating periodontal diseases therefore implies action against microorganisms established on tooth surfaces. Contemporary periodontal treatment consists principally of meticulous mechanical cleaning of tooth surfaces to remove bacterial deposits (plaque and calculus). Since mechanical procedures alone cannot

Table 1. Comparison of local and systemic antimicrobial therapy

Issue	Systemic administration	Local administration
Drug distribution	wide	narrow effective range
Drug concentration	variable levels in different body compartments	high dose at treated site, low levels elsewhere
Therapeutic potential	may reach widely distributed microorganisms better	may act locally on biofilm-associated bacteria better
Problems	systemic side effects	reinfection from nontreated sites
Clinical limitations	requires good patient compliance	requires infection to be limited to the treated site

eliminate the bacteria involved completely, various protocols for adjunctive use of antibacterial agents have been devised. Antibiotics are an efficient antibacterial means of treating bacterial infections and, therefore, merit thorough evaluation in the management of periodontal disease. Early reports noted beneficial side effects in the periodontal condition of patients that had been treated with antibiotics for other reasons. One example is the article published in *The Lancet* in 1962 that reported relief from 'acute marginal gingivitis' in a female patient, originally treated for a trichomonal vaginitis with metronidazole [1]. This paper marks the starting point for using imidazole compounds in periodontics. In the 5 decades since that publication, a large number of well-designed clinical trials have been conducted, showing beneficial effects of antimicrobial protocols in the treatment of periodontal diseases. On the basis of these trials, but to a large extent also on extrapolations from other research, deductive reasoning and plausibility, recommendations for the proper use of antibiotics in periodontics have been issued. Several questions have been raised and warnings have been articulated against the unrestricted use of antibiotics in treating periodontal diseases. Some of these questions have been resolved for good; others remain a matter of dispute.

This paper will use the latest evidence to address various paradigms and questions in the context of antimicrobial treatment of periodontal diseases and will indicate likely future directions.

What Options Do We Have?

Oral antibiotics are the most common way of treating bacterial infections. Using appropriate delivery devices, local manifestations can also be addressed with agents that are directly administered into infected sites. Table 1 compares the potentials and limitations of local and systemic antimicrobial therapy. Local therapy may allow the application of drugs at levels that cannot be reached by the systemic route and may be suitable for agents, i.e. antiseptics, that are too toxic to be delivered by the systemic route. Local therapy presupposes that the infection is limited to the clinically diagnosed lesion. Bacteria typically associated with periodontal lesions can, however, also be detected in clinically unaffected regions of the dentition [2, 3] and non-dental sites, such as the dorsum of the tongue or tonsillar crypts [4–8]. Systemic adverse drug reactions, on the other hand, are a greater concern and more likely to occur if drugs are distributed via the systemic route. Even mild secondary reactions may severely decrease patient compliance [9], whereas local delivery is independent of patient compliance.

A restricted number of antimicrobial agents have been thoroughly tested for use in periodontal therapy. The drugs more extensively investigated for systemic use include amoxicillin (with or without clavulanic acid), azithromycin, clindamycin, doxycycline, metronidazole, spiramycin, tetracycline and certain combinations thereof [10]. The drugs tested thoroughly for local application commonly include tetracycline, minocycline, doxycycline, metronidazole and chlorhexidine.

Among the penicillins, amoxicillin has been favored for treatment of periodontal disease because of its considerable activity against several periodontal pathogens at levels available in gingival fluid. The molecular structure of penicillins includes a β-lactam ring that may be cleaved by bacterial enzymes. Clavulanic acid can been added to amoxicillin to inhibit bacterial β-lactamase activity. The tetracyclines clindamycin and erythromycin have a broad spectrum of activity and are bacteriostatic. In addition to their antimicrobial effect, tetracyclines are capable of inhibiting collagenase [11]. The nitroimidazoles (metronidazole and ornidazole) and the quinolone antibiotics (e.g. ciprofloxacin) act by inhibiting DNA synthesis. Metronidazole acts specifically against the obligately anaerobic microbiota, which includes many but not all suspected periodontal pathogens.

Some combinations of antibiotics may offer an enlarged spectrum of activity and may induce less bacterial resistance than the single agents. Combination drug therapy may be especially useful when acting against a complex subgingival microbiota that might harbor multiple pathogenic species with different antimicrobial susceptibility. A synergistic effect against *Aggregatibacter actinomycetemcomitans* has been noted in vitro between metronidazole and its hydroxy metabolite [12, 13] as well as between these two compounds and amoxicillin [14].

Is Periodontitis an Infection and Should We Treat It as One?

The term infection refers to the presence and multiplication of microorganisms in a host organism, whereby the infecting agent usually invades body tissues. The specific conditions for antimicrobial therapy of periodontal diseases are shown in figure 1.

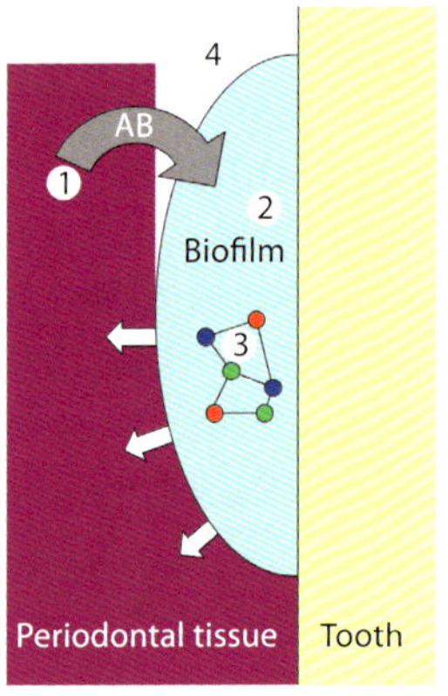

Fig. 1. Specific conditions for the use of antimicrobial agents in periodontal therapy. Microorganisms are located outside of the diseased tissues. Agents must reach periodontal pocket environment (1). Microorganisms form a biofilm on the root surface (2). Therapy is targeted at a complex microbiota (3). The periodontal pocket as an open site is subject to recolonization after therapy (4).

Although there is evidence of bacterial penetration in acute periodontal conditions, notably in periodontal abscesses and ulcerative lesions [15–19], it has not been established that periodontal disease results from bacterial invasion and the multiplication of bacteria within tissues. The uniqueness of plaque-associated dental diseases as infections rather relates to the lack of massive bacterial tissue invasion. Therefore, it is believed that in order to be successful, any antimicrobial protocol should aim at the subgingival microbiota by delivering the active principle at a therapeutic dose into the periodontal pocket environment.

The microbiota that accumulate on tooth surfaces form a mixed sessile community of cells, attached to the surface and to each other, embedded in a matrix of extracellular polymeric substances that they have produced, termed a biofilm. A discussion of the biofilm phenomenon is provided in the chapters by Zijinge et al. [pp. 1–16] and Diaz [pp. 17–40] and will not be repeated here. Comprehensive review papers have highlighted the importance of biofilm formation for understanding the etiology of periodontal diseases and therapeutic consequences in a broader perspective [20–23]. In brief, biofilm-associated infections are notoriously resistant to antimicrobial therapy unless the biofilm is disrupted mechanically. Multiple factors appear to contribute to the overall resistance of biofilm bacteria. These include the protection offered by extracellular polymeric substances leading to the failure of the antimicrobial agent to penetrate the biofilm, and the adoption of a resistant physiological state or phenotype related to the multicellular nature of the biofilm community. Biofilms play an important role in the spread of antibiotic resistance. Within the dense bacterial population, efficient horizontal transfer of resistance and virulence genes takes place. In recognition of these mechanisms, there is general consensus today that antimicrobial therapy should be preceded by thorough mechanical debridement to disrupt the structured bacterial aggregates.

Is There Microbial Specificity and Should Specific Pathogens Be Targeted?

Treatment of periodontal diseases targets a highly diverse microbiota [24, 25]. It has been suggested that among the numerous recognized taxa, only a few gram-negative bacterial species play a specific role [26]. For two organisms, *Porphyromonas gingivalis* and *A. actinomycetemcomitans*, longitudinal or retrospective studies have indicated an increased risk of periodontal breakdown in positive sites and better treatment outcomes where organisms could no longer be detected at follow-up [27–35]. *A. actinomycetemcomitans* displays a broad genetic and phenotypic diversity and is heterogeneously distributed in various populations and cohorts worldwide [36]. In one large prospective study [37], only one subpopulation of *A. actinomycetemcomitans*, the 'JP2 clone' [38], showed the properties of a true pathogen. There is an ongoing debate on the utility of microbiological tests to identify such organisms in order to optimize periodontal therapy.

The combination of amoxicillin plus metronidazole has a proven capacity to suppress *A. actinomycetemcomitans* from periodontitis lesions and other oral sites [6, 14, 39–45] and is therefore the first choice of many clinicians, especially for the treatment of advanced *A.-actinomycetemcomitans*-associated periodontitis. However, is there evidence to limit this therapy to patients testing positive for *A. actinomycetemcomitans* – perhaps even to only those testing 'JP2' positive. Somewhat surprisingly, the data available from randomized clinical trials was until recently inconclusive due to small numbers of *A.-actinomycetemcomitans*-positive subjects treated with antibiotics. Four randomized clinical trials can be cited [46–49] that include *A.-actinomycetemcomitans*-positive and negative patients treated with or without adjunctive systemic amoxicillin and metronidazole. Although *A. actinomycetemcomitans* could no longer be detected after antibiotic therapy in all subjects of these 4 studies, a clear-cut specific clinical benefit only emerging in *A. actinomycetemcomitans*-positive subjects was not demonstrated. To the contrary however, in two trials [48, 49] those *A.-actinomycetemcomitans*-negative subjects that received the antibiotics had a significantly better clinical outcome than those that were treated with placebo.

Here, the results of an intermediate analysis from a double-blind placebo-controlled randomized longitudinal study – including 28 *A.-actinomycetemcomitans*-positive and 41 *A.-actinomycetemcomitans*-negative subjects – are described [50]. All 69 patients received full-mouth periodontal debridement, performed within 48 h; 34 subjects then received systemic antibiotics for 7 days (500 mg metronidazole + 375 mg amoxicillin three times per day), while 35 subjects received a placebo. Probing depth and bleeding on probing were monitored at 6 sites on all teeth (5,365 sites in total) before and 3 months after treatment. Persistence of sites with a probing depth >4 mm and bleeding on probing (G4B) was the primary outcome variable. Using multilevel logistic regression, the effect of antibiotics on a positive G4B at month 3 was analyzed according to the following factors (interaction effect): *A. actinomycetemcomitans* positive or negative at baseline, gender, age, smoking, tooth being a molar and interdental location. After treatment, test subjects had a significantly lower individual mean G4B than control subjects (p = 0.001). Being *A. actinomycetemcomitans* positive or negative did not change the effect of the antibiotics. Subjects benefited from the antibiotics irrespectively of gender, age or smoking status. It was therefore concluded that even though amoxicillin plus metronidazole very efficiently suppressed *A. actinomycetemcomitans,* testing for *A. actinomycetemcomitans* among subjects with a clinical diagnosis of chronic periodontitis did not identify subjects who would specifically benefit from antibiotics. Furthermore, it must be stated that until an interventional study identifies a specific advantage of a treatment other than amoxicillin plus metronidazole in JP2-positive cases, the utility of typing *A. actinomycetemcomitans* remains hypothetical.

It has been argued that the administration of antibiotics to *P.-gingivalis*-negative patients may be overtreatment [51]. The authors who raised this issue cited a study of their own where 49 subjects with chronic periodontitis were treated with full-mouth scaling and root planing, plus amoxicillin and metronidazole or placebo [47]. In this study, *P.-gingivalis*-positive patients showed significantly better results when treated with antibiotics than when treated with placebo. Interestingly, a closer look at their data however shows that the percentage of pockets >4 mm also decreased very similarly in *P.-gingivalis*-negative subjects treated with antibiotics (from 40 to 12% in *P.-gingivalis*-negative subjects; from 46 to 11% in *P.-gingivalis*-positive subjects). At least 3 other studies [46, 48, 49] have clearly demonstrated some benefit in supplementing mechanical therapy with amoxicillin and metronidazole in *P.-gingivalis*-negative cases. Furthermore, even if treatment of *P.-gingivalis*-negative patients with antibiotics were overtreatment, one could point out that this concerns only a small group of patients seeking treatment for periodontitis. In 182 subjects with untreated periodontitis, who sought periodontal treatment at the School of Dental Medicine at the University of Geneva, *P. gingivalis* was frequently detected (89%). Based on a clinical diagnosis of untreated periodontitis, positive outcomes of microbiological tests for *P. gingivalis,* as well as for two other periodontal marker organisms *(T. forsythia* and *T. denticola)* could be predicted with high confidence, irrespective of a patient's origin, smoking status or age (unpublished data).

It has been stated that 'systemic antibiotic therapy in periodontics aims to eliminate or markedly suppress specific microorganisms with the potential to cause breakdown of periodontal attachment in susceptible patients' [52]. However, it has not been proven that the selective suppression of specific members of the subgingival microbial complex is the key element for success. Given the large diversity of the microbiota associated with all forms of periodontitis and multiple synergistic and antagonistic interactions among the members of the flora, the concept of specifically identifying and eradicating a particular pathogen may be illusionary. Since the periodontal flora never consists of one single species, synergistic, but also antagonistic, relationships between microorganisms could occur.

Table 2. Potential benefits and risks of early or late adjunctive antimicrobial therapy

Administration		Effects
Early (initial nonsurgical treatment phase)	benefits	shorter anti-infective treatment; reduced need for periodontal surgery
	risks	incomplete removal of subgingival biofilm and calculus at time of administration
Late (secondary treatment phase)	benefits	constrains prescription of antibiotics to cases that cannot be resolved by scaling and root planing alone
	risks	no evidence of advantage from specifically designed clinical trials; benefits of antibiotics used as adjunct to surgical procedures unclear

Which Is the Optimal Time Point for Administering Antibiotics?

The optimal timing of antimicrobial drug administration has been a subject of controversy for many years (table 2). The first study demonstrating a reduced need for surgical therapy in patients that were treated early with systemic metronidazole as an adjunct to scaling and root planing [53], as well as a later publication claiming sustained benefits of initial antimicrobial therapy after 5 years [54], met with disapproval from prominent members of the periodontal community. Postponing antibiotic therapy to the second surgical treatment phase may be defended using two arguments. First, as it is known that scaling and root planing alone is able to resolve a considerable amount of periodontal pathology on its own [55, 56], this strategy may help to constrain the prescription of antibiotics to a minimum. Second, given the restricted effects of antibiotics on intact biofilm [57], and the known limitations of non-surgical mechanical debridement [58, 59], a surgical intervention may be needed for access to ensure complete removal of subgingival biofilm and calculus. As much as this reasoning seems to make sense, data from specifically designed controlled trials are unavailable to support this recommendation. Contrary to this view, the vast majority of trials have tested antibiotics as adjuncts to non-surgical debridement. A systematic review in 2008 trying to assess the relative benefit of prescribing antibiotics either during the non-surgical or the surgical phase of therapy remained inconclusive [60]. However, two studies, not included in this review, showed better clinical outcomes for subjects with generalized aggressive periodontitis if patients were given amoxicillin and metronidazole immediately after initial scaling and root planing rather than after re-treatment of persisting pathology [61, 62].

Are There Specific Benefits in Certain Situations?

In 51 patients diagnosed with chronic periodontitis, systemic metronidazole and amoxicillin significantly improved the 6-month clinical outcomes of extensive full-mouth non-surgical periodontal debridement provided in 48 h, thereby significantly reducing the need for additional therapy [49, 63]. In a trial involving 25 patients, significantly better clinical results of full-mouth debridement resulted after using an ultrasonic scaler for only 45 min when this intervention was supplemented with adjunctive amoxicillin plus metronidazole [64]. A third study [65] showed the adjunctive benefit of the same antibiotic regime after full-mouth debridement in 41 patients with generalized aggressive periodontitis. Taken together, the data from these trials illustrate the robustness and reliability of the beneficial effect of adjunctive amoxicillin and metronidazole, independent of diagnosis and varied treatment protocols.

Based on the concept that the presence of beneficial bacterial species may suppress the activity of pathogens, one can speculate that it may be advantageous to specifically eliminate target bacteria only and to allow the growth of potentially beneficial microorganisms. Such contemplations have been used as an argument to propagate narrow-spectrum antibiotics for periodontal therapy. However, in 2 comparative studies, better clinical outcomes were reported if patients were treated with amoxicillin and metronidazole than if metronidazole alone was administered [48, 66]. At present, no randomized clinical trial has been able to demonstrate superiority of any other regime over amoxicillin and metronidazole in any clinically or microbiologically defined variant of periodontal disease.

What Is the Optimal Administration Pathway?

Apart from systemic administration, various methods to deliver antimicrobial agents into periodontal pockets have been devised and studied. They include pocket irrigation, the placement of drug-containing ointments and gels, and devices for sustained drug release.

A systematic review [67] analyzing the literature in an effort to determine the effect of subgingival irrigation concluded that irrigation with chlorhexidine provided no additional benefit to conventional manual instrumentation. Another systematic review found no significant benefit from supplementing full-mouth debridement with chlorhexidine irrigation [68]. A recent systematic review specifically addressed rinsing with povidone-iodine during nonsurgical periodontal therapy [69]. A small but statistically significant beneficial effect of pocket rinsing in terms of probing depth reduction was found.

Local antimicrobial therapy may specifically be suited for residual pockets. To control the regrowth of a microbiota incompatible with periodontal health [70–72], such sites require specific professional attention and are subject to frequent instrumentation. Systemic antibiotics are not ideal for situations necessitating repeated intervention. The efficacy of several commercially available delivery systems for local antimicrobial therapy has been tested in clinical trials. Two systematic reviews have tried to evaluate the combined literature-based evidence to determine their relative effect [73, 74]. Due to the heterogeneity of the material, the authors could not make any firm statements regarding the superiority of any one system. Findings must be interpreted with caution as differences in the outcomes may reflect dissimilarity in modes of application and study populations, rather than the potencies of the agents. In recent years, the perceived value of local antibiotic therapy has decreased, as the majority of adequately tested formulations have been withdrawn from the market. A recent cost-effectiveness analysis concluded that systemic antimicrobials are more cost-effective than locally delivered antimicrobials [75]. So far, no local antimicrobial treatment has proven to be equally efficient or better than systemic amoxicillin and metronidazole. One of the few studies truly evaluating this possibility concluded that the adjunctive placement of chlorhexidine chips was less efficacious in the treatment of aggressive periodontitis [76].

As there is a need to carefully evaluate new protocols that are safe and effective for maintenance, we investigated the effects on residual pockets of treatment with a diode laser and photodynamic therapy (PDT). The diode laser with a wavelength between 655 and 980 does not interact with dental hard tissues. It is used for cutting and coagulating soft tissue, and has been proposed for sulcular debridement and curettage. PDT is based on the principle that light of a suitable wavelength can activate certain substances, called 'photosensitizers', to produce free oxygen radicals able to destroy bacteria and their products. Both methods may have a bactericidal and detoxifying effect, and may therefore be useful as an adjunct or alternative to conventional mechanical instrumentation of residual pockets. Debridement with an ultrasonic device, followed by either PDT or diode soft laser application, decreased pocket depths of residual pockets and their tendency to bleed upon probing (unpublished data). Fewer pockets >4 mm and reduced bleeding on probing remained 6 months after PDT or conventional instrumentation with sharp instruments than after diode soft laser application. Two systematic reviews evaluated the accumulated evidence regarding the potential benefit of laser application compared with mechanical debridement in non-surgical periodontal therapy [77, 78]. Due to the heterogeneity of studies, a meta-analysis could not be performed, and the evidence to support the clinical application of different diode laser wavelengths was inconclusive. The effectiveness of PDT, as the sole mode of primary non-surgical periodontal therapy or an adjunct to scaling and root planing, was recently evaluated in a systematic review [79]. Five studies could be evaluated. PDT, as an independent treatment or as an adjunct to scaling and root planing, did not demonstrate a clear advantage over scaling and root planing alone. The authors concluded that more research is necessary for a proper evaluation of this therapy.

What Is the Balance of Benefits and Risks for Antibiotics?

Amoxicillin and metronidazole have been widely prescribed for over 3 decades, and their effects and side effects are therefore well documented [80, 81]. The most frequent adverse reactions to amoxicillin are sensitivity phenomena. These are more likely to occur in individuals who have previously demonstrated hypersensitivity to penicillins and in those with a history of allergies in general. Most reactions are mild and limited to a rash or skin lesion in the head or neck region. More severe reactions may induce swelling and tenderness of joints. In highly sensitized patients, a life-threatening anaphylactic reaction may develop [81]. Potential side effects of metronidazole include nausea, headaches, loss of appetite, diarrhea, a metallic taste and rarely a rash. Alcohol consumption enhances these symptoms, because imidazoles affect the activity of liver enzymes. Peripheral neuropathies, characterized mainly by numbness or paresthesia of an extremity, have been reported from isolated cases. Therefore, metronidazole should be discontinued immediately if abnormal neurological signs appear [80]. Metronidazole has shown evidence of carcinogenic activity in studies involving chronic oral administration in mice and rats, but not in other tested species. Due to inadequate evidence, metronidazole is not considered as a risk factor for cancer in humans [82]. Previously unrecognized candidiasis may present more prominent symptoms during antibiotic therapy.

To put things into perspective, the frequency and consequences of adverse effects of antibiotics have to be balanced against potential health consequences of not rapidly suppressing a periodontal infection, and with the inconvenience, discomfort and financial consequences of further therapy. The traditional approach sometimes expands treatment over several months, while scaling and root planing plus amoxicillin and metronidazole may be able to resolve the infection within a few days. Scaling and root planing plus amoxicillin and metronidazole have been shown to decrease clinical signs of inflammation and inflammatory biomarkers in gingival crevice fluid more profoundly than scaling and root planing alone [83]. Although not directly confirmed so far by a clinical trial, it seems preferable, from a general health point of view, to let patients benefit early from the positive systemic effects of successful periodontal therapy [84–86]. Although the number of subjects complaining about gastrointestinal problems, notably diarrhea, was indeed higher in the test group than the placebo group in a clinical trial [63], it is also worth considering that tooth loss and suppuration despite therapy were exclusively noted in the control group.

Can We Minimize the Risk of the Development of Microbial Antibiotic Resistance?

Antibiotic therapy carries the risk of promoting the development of bacterial antibiotic resistance. Since periodontal diseases are not life-threatening and can be managed by and large without antibiotics, this concern needs to be taken seriously. However,

Table 3. Strategies to reduce the risk of bacterial antimicrobial resistance

Issues	Recommendations
Agents	consider prescription of a combination therapy (i.e. amoxicillin + metronidazole)
Dosage and timing	administer a high dose for a short period
Clinical protocol	use antimicrobials as adjuncts to thorough mechanical debridement
Indication	use antibiotics for therapy where scaling and root planing alone are insufficient; restrict prophylactic use to high-risk patients and limit use to prevent severe complications

the impact of various treatment protocols on the development of bacterial antibiotic resistance has unfortunately not been evaluated thus far in specifically designed clinical trials. Proposed strategies to reduce the risk of bacterial antimicrobial resistance are listed in table 3. In the mixed subgingival microflora, some microorganisms can always be expected to be resistant to any single antimicrobial agent [87]. Therefore, to overcome single antibiotic resistance, combination therapies may be advantageous.

It has been recognized that the dose and duration of antimicrobial therapy are crucial parameters for resistance development. Clinical studies in other fields have indicated that it is preferable to deliver antibiotics at a high dose over a short time. As an example, a randomized trial conducted in an outpatient clinic involving 795 children receiving antibiotic prescriptions for respiratory tract illness demonstrated the advantage of short-course high-dose outpatient antibiotic therapy to minimize the impact of antibiotic use on the spread of drug-resistant pneumococci [88]. Similar studies are currently unavailable for periodontal treatments. As discussed above, biofilm-associated infections are notoriously resistant in general to antimicrobial therapy unless the biofilm is disrupted mechanically. It has therefore been advised to precede all antimicrobial therapy by mechanical debridement whenever such an approach is possible.

It is currently unknown to what extent a dentist treating a periodontal infection with antibiotics contributes to the resistance problem overall, especially if he uses two drugs such as amoxicillin and metronidazole together only once during the lifetime of his patient and in a controlled situation following thorough mechanical debridement. This contribution may be minor compared to the effects of the frequent prescription of antibiotics by various healthcare providers for other therapeutic and prophylactic purposes. Antibiotics are taken by thousands of subjects with untreated periodontitis worldwide every day without subgingival debridement. This happens whenever antibiotics are prescribed for medical reasons to persons with unrecognized or untreated periodontal disease. As an example, in a cohort of over 12,000 persons with destructive periodontal disease, more than 70% received at least one course of antibiotics over a 3-year period

[89]. A treatment protocol implementing the current evidence has been presented recently [90]. Mild periodontitis can, and should, be treated non-surgically, and without antibiotics. Cases with multiple deep pockets should first receive a thorough mechanical debridement before antibiotics are administered. Prophylactic use should be restricted to high-risk patients and limited in order to prevent severe complications [91–95].

Are Further Advances Possible?

Numerous antimicrobial agents, including novel derivatives of existing antibiotic classes (such as the quinolones, macrolides/ketolides, tetracyclines, streptogramins, glycopeptides and β-lactams), are currently being investigated. Compounds that act on unique targets are the subject of specific discovery programs [96]. Many preclinical investigations focus on action against antibiotic-resistant medical pathogens [97]. Efforts are being made to develop antimicrobial peptides with bactericidal properties to kill both gram-positive and gram-negative organisms, and this topic is discussed in the chapter by Gorr [pp. 84–98]. Such agents have shown activity against clinically important multi-resistant bacteria, such as vancomycin-resistant enterococci and methicillin-resistant *Staphylococcus aureus* [98]. However, an extensive discussion of new antibiotic developments and discoveries from a pharmaceutical point of view is beyond the scope of this review.

In light of the excellent clinical results of combination therapy with well-established drugs that are cheap and efficient, newly proposed protocols for periodontal therapy will have to be compared to positive controls in clinical trials, not only to a placebo. Microbiological evaluations will have to focus not only on action against microorganisms directly involved in periodontal diseases, i.e. the predominantly gram-negative complex of the subgingival microbiota, but also on organisms relevant for other reasons, such as staphylococci or enterococci.

Acknowledgement

The author's work on antimicrobial therapy as a adjunct to periodontal treatment is supported by the Swiss National Science Foundation (grant No. 320030-122089).

References

1 Shinn DLS: Metronidazole in acute ulcerative gingivitis. Lancet 1962;1191.

2 Mombelli A, McNabb H, Lang NP: Black-pigmenting gram-negative bacteria in periodontal disease. I. Topographic distribution in the human dentition. J Periodontal Res 1991;26:301–307.

3 Mombelli A, Gmur R, Gobbi C, Lang NP: *Actinobacillus actinomycetemcomitans* in adult periodontitis. I. Topographic distribution before and after treatment. J Periodontol 1994;65:820–826.

4 Müller HP, Eickholz P, Heinecke A, Pohl S, Müller RF, Lange DE: Simultaneous isolation of *Actinobacillus actinomycetemcomitans* from subgingival and extracrevicular locations of the mouth. J Clin Periodontol 1995;22:413–419.

5 Müller HP, Lange DE, Müller RF: Failure of adjunctive minocycline-HCl to eliminate oral *Actinobacillus actinomycetemcomitans*. J Clin Periodontol 1993;20: 498–504.

6 Pavicic MJAMP, van Winkelhoff AJ, Douqué NH, Steures RWR, de Graaff J: Microbiological and clinical effects of metronidazole and amoxicillin in *Actinobacillus actinomycetemcomitans*-associated periodontitis. J Clin Periodontol 1994;21:107–112.

7 van Winkelhoff AJ, van der Velden U, Clement M, de Graaff J: Intra-oral distribution of black-pigmented *Bacteroides* species in periodontitis patients. Oral Microbiol Immunol 1988;3:83–85.

8 Zambon JJ, Reynolds HS, Slots J: Black-pigmented *Bacteroides* spp. in the human oral cavity. Infect Immun 1981;32:198–203.

9 Loesche WJ, Grossman N, Giordano J: Metronidazole in periodontitis (IV). The effect of patient compliance on treatment parameters. J Clin Periodontol 1993;20:96–104.

10 Slots J, Greenwell H, Fiorellini J, Giannobile W, Offenbacher S, Salkin L: Systemic antibiotics in periodontics. J Periodontol 2004;75:1553–1565.

11 Golub LM, Wolff M, Lee HM, McNamara TF, Ramamurthy NS, Zambon J, Ciancio S: Further evidence that tetracyclines inhibit collagenase activity in human crevicular fluid and other mammalian sources. J Periodontal Res 1985;20:12–23.

12 Jousimies-Somer H, Asikainen S, Suomala P, Summanen P: Activity of metronidazole and its hydroxymetabolite. Oral Microbiol Immunol 1988; 3:32–34.

13 Pavicic MJAMP, van Winkelhoff AJ, de Graaff J: Synergistic effects between amoxicillin, metronidazole, and the hydroxymetabolite of metronidazole against *Actinobacillus actinomycetemcomitans*. Antimicrob Agents Chemother 1991;35:961–966.

14 Pavicic MJAMP, van Winkelhoff AJ, de Graaff J: In vitro susceptibilities of *Actinobacillus actinomycetemcomitans* to a number of antimicrobial combinations. Antimicrob Agents Chemother 1992;36: 2634–2638.

15 Allenspach-Petrzilka GE, Guggenheim B: Bacterial invasion of the periodontium: an important factor in the pathogenesis of periodontitis? J Clin Periodontol 1983;10:609–617.

16 Saglie R, Newman MG, Carranza FA, Pattison GL: Bacterial invasion of gingiva in advanced periodontitis in humans. J Periodontol 1982;53:217–222.

17 Saglie FR, Carranza FA Jr, Newman MG, Cheng L, Lewin KJ: Identification of tissue-invading bacteria in human periodontal disease. J Periodontal Res 1982;17:452–455.

18 Carranza FA Jr, Saglie R, Newman MG, Valentin PL: Scanning and transmission electron microscopic study of tissue-invading microorganisms in localized juvenile periodontitis. J Periodontol 1983;54: 598–617.

19 Listgarten MA: Electron microscopic observations on the bacterial flora of acute necrotizing ulcerative gingivitis. J Periodontol 1965;36:328–339.

20 Costerton JW, Montanaro L, Arciola CR: Biofilm in implant infections: its production and regulation. Int J Artif Organs 2005;28:1062–1068.

21 Costerton JW: Biofilm theory can guide the treatment of device-related orthopaedic infections. Clin Orthop Relat Res 2005;437:7–11.

22 Davey ME, Costerton JW: Molecular genetics analyses of biofilm formation in oral isolates. Periodontol 2000 2006;42:13–26.

23 Marsh PD, Moter A, Devine DA: Dental plaque biofilms: communities, conflict and control. Periodontol 2000 2011;55:16–35.

24 Paster BJ, Boches SK, Galvin JL, Ericson RE, Lau CN, Levanos VA, Sahasrabudhe A, Dewhirst FE: Bacterial diversity in human subgingival plaque. J Bacteriol 2001;183:3770–3783.

25 Kroes I, Lepp PW, Relman DA: Bacterial diversity within the human subgingival crevice. Proc Natl Acad Sci USA 1999;96:14547–14552.

26 Socransky SS, Haffajee AD: Microbial mechanisms in the pathogenesis of destructive periodontal diseases: a critical assessment. J Periodontal Res 1991; 26:195–212.

27 Bragd L, Dahlén G, Wikström M, Slots J: The capability of *Actinobacillus actinomycetemcomitans*, *Bacteroides gingivalis* and *Bacteroides intermedius* to indicate progressive periodontitis; a retrospective study. J Clin Periodontol 1987;14:95–99.

28 Grossi SG, Zambon JJ, Ho AW, Koch G, Dunford RG, Machtei EE, Norderyd OM, Genco RJ: Assessment of risk for periodontal disease. I. Risk indicators for attachment loss. J Periodontol 1994; 65:260–267.

29 Haffajee AD, Socransky SS: Microbial etiological agents of destructive periodontal diseases. Periodontol 2000 1994;5:78–111.

30 Haffajee AD, Socransky SS, Smith C, Dibart S: Relation of baseline microbial parameters to future periodontal attachment loss. J Clin Periodontol 1991;18:744–750.

31 Carlos JP, Wolfe MD, Zambon JJ, Kingman A: Periodontal disease in adolescents: some clinical and microbiologic correlates of attachment loss. J Dent Res 1988;67:1510–1514.

32 Chaves ES, Jeffcoat MK, Ryerson CC, Snyder B: Persistent bacterial colonization of *Porphyromonas gingivalis*, *Prevotella intermedia*, and *Actinobacillus actinomycetemcomitans* in periodontitis and its association with alveolar bone loss after 6 months of therapy. J Clin Periodontol 2000;27:897–903.

33 Rams TE, Listgarten MA, Slots J: The utility of 5 major putative periodontal pathogens and selected clinical parameters to predict periodontal breakdown in adults on maintenance care. J Clin Periodontol 1996;23:346–354.

34 Dahlén G, Wikström M, Renvert S: Treatment of periodontal disease based on microbiological diagnosis: a 5-year follow-up on individual patterns. J Periodontol 1996;67:879–887.

35 Wennström JL, Dahlén G, Svensson J, Nyman S: *Actinobacillus actinomycetemcomitans*, *Bacteroides gingivalis* and *Bacteroides intermedius*: predictors of attachment loss? Oral Microbiol Immunol 1987;2: 158–163.

36 Kilian M, Frandsen EV, Haubek D, Poulsen K: The etiology of periodontal disease revisited by population genetic analysis. Periodontol 2000 2006;42:158–179.

37 Haubek D, Ennibi OK, Poulsen K, Vaeth M, Poulsen S, Kilian M: Risk of aggressive periodontitis in adolescent carriers of the JP2 clone of *Aggregatibacter (Actinobacillus) actinomycetemcomitans* in Morocco: a prospective longitudinal cohort study. Lancet 2008;371:237–242.

38 Tsai CC, Shenker BJ, DiRienzo JM, Malmud D, Taichman NS: Extraction and isolation of a leukotoxin from *Actinobacillus actinomycetemcomitans* with polymyxin B. Infect Immun 1984;43:700–705.

39 Goené RJ, Winkel EG, Abbas F, Rodenburg JP, Van Winkelhoff AJ, De Graaff J: Microbiology in diagnosis and treatment of severe periodontitis: a report of four cases. J Periodontol 1990;61:61–64.

40 van Winkelhoff AJ, Rodenburg JP, Goené RJ, Abbas F, Winkel EG, de Graaff J: Metronidazole plus amoxicillin in the treatment of *Actinobacillus actinomycetemcomitans* associated periodontitis. J Clin Periodontol 1989;16:128–131.

41 van Winkelhoff AJ, Tijhof CJ, de Graaff J: Microbiological and clinical results of metronidazole plus amoxicillin therapy in *Actinobacillus actinomycetemcomitans*-associated periodontitis. J Periodontol 1992;63:52–57.

42 Berglundh T, Krok L, Liljenberg B, Westfelt E, Serino G, Lindhe J: The use of metronidazole and amoxicillin in the treatment of advanced periodontal disease: a prospective, controlled clinical trial. J Clin Periodontol 1998;25:354–362.

43 Flemmig TF, Milian E, Kopp C, Karch H, Klaiber B: Differential effects of systemic metronidazole and amoxicillin on *Actinobacillus actinomycetemcomitans* and *Porphyromonas gingivalis* in intraoral habitats. J Clin Periodontol 1998;25:1–10.

44 Dannewitz B, Pohl S, Eickholz P, Kim TS: Clinical and microbiological effects of a combined mechanic-antibiotic therapy in subjects with *Actinobacillus actinomycetemcomitans*-associated periodontitis. Am J Dent 2007;20:153–156.

45 Mombelli A, Schmid B, Rutar A, Lang NP: Local antibiotic therapy guided by microbiological diagnosis: treatment of *Porphyromonas gingivalis* and *Actinobacillus actinomycetemcomitans* persisting after mechanical therapy. J Clin Periodontol 2002; 29:743–749.

46 Flemmig TF, Milian E, Karch H, Klaiber B: Differential clinical treatment outcome after systemic metronidazole and amoxicillin in patients harboring *Actinobacillus actinomycetemcomitans* and/or *Porphyromonas gingivalis*. J Clin Periodontol 1998;25:380–387.

47 Winkel EG, Van Winkelhoff AJ, Timmerman MF, Van der Velden U, Van der Weijden GA: Amoxicillin plus metronidazole in the treatment of adult periodontitis patients: a double-blind placebo-controlled study. J Clin Periodontol 2001;28:296–305.

48 Rooney J, Wade WG, Sprague SV, Newcombe RG, Addy M: Adjunctive effects to non-surgical therapy of systemic metronidazole and amoxycillin alone and combined. A placebo controlled study. J Clin Periodontol 2002;29:342–350.

49 Cionca N, Giannopoulou C, Ugolotti G, and Mombelli A: Microbiologic testing and outcomes of full-mouth scaling and root planing with or without amoxicillin/metronidazole in chronic periodontitis. J Periodontol 2010;81:15–23.

50 Mombelli A, Cionca N, Almaghlouth A, Décaillet F, Courvoisier D, Giannopoulou E: Are there specific benefits of amoxicillin-metronidazole in AA-positive periodontitis? J Dent Res 2011;90(spec iss B):1757.

51 van Winkelhoff AJ, Winkel EG: Antibiotics in periodontics: right or wrong? J Periodontol 2009;80: 1555–1558.

52 van Winkelhoff AJ, Rams TE, Slots J: Systemic antibiotic therapy in periodontics. Periodontol 2000 1996;10:45–78.

53 Loesche WJ, Giordano JR, Hujoel P, Schwarcz J, Smith BA: Metronidazole in periodontitis: reduced need for surgery. J Clin Periodontol 1992;19:103–112.
54 Loesche WJ, Giordano JR, Soehren S, Kaciroti N: The nonsurgical treatment of patients with periodontal disease: results after five years. J Am Dent Assoc 2002;133:311–320.
55 van der Weijden GA, Timmerman FA: A systematic review on the clinical efficacy of subgingival debridement in the treatment of chronic periodontitis. J Clin Periodontol 2002;29(suppl 3):55–71.
56 Heitz-Mayfield LJ, Trombelli L, Heitz F, Needleman I, Moles D: A systematic review of the effect of surgical debridement vs non-surgical debridement for the treatment of chronic periodontitis. J Clin Periodontol 2002;29(suppl 3):92–102.
57 Sedlacek MJ, Walker C: Antibiotic resistance in an in vitro subgingival biofilm model. Oral Microbiol Immunol 2007;22:333–339.
58 Rabbani GM, Ash MM Jr, Caffesse RG: The effectiveness of subgingival scaling and root planing in calculus removal. J Periodontol 1981;52:119–123.
59 Buchanan SA, Robertson PB: Calculus removal by scaling/root planing with and without surgical access. J Periodontol 1987;58:159–163.
60 Herrera D, Alonso B, Leon R, Roldan S, Sanz M: Antimicrobial therapy in periodontitis: the use of systemic antimicrobials against the subgingival biofilm. J Clin Periodontol 2008;35:45–66.
61 Kaner D, Christan C, Dietrich T, Bernimoulin JP, Kleber BM, Friedmann A: Timing affects the clinical outcome of adjunctive systemic antibiotic therapy for generalized aggressive periodontitis. J Periodontol 2007;78:1201–1208.
62 Griffiths GS, Ayob R, Guerrero A, Nibali L, Suvan J, Moles DR, Tonetti MS: Amoxicillin and metronidazole as an adjunctive treatment in generalized aggressive periodontitis at initial therapy or re-treatment: a randomized controlled clinical trial. J Clin Periodontol 2011;38:43–49.
63 Cionca N, Giannopoulou C, Ugolotti G, Mombelli A: Amoxicillin and metronidazole as an adjunct to full-mouth scaling and root planing of chronic periodontitis. J Periodontol 2009;80:364–371.
64 Ribeiro Del Peloso E, Bittencourt S, Zanin IC, Bovi Ambrosano GM, Sallum EA, Nociti FH, Goncalves RB, Casati MZ: Full-mouth ultrasonic debridement associated with amoxicillin and metronidazole in the treatment of severe chronic periodontitis. J Periodontol 2009;80:1254–1264.
65 Guerrero A, Griffiths GS, Nibali L, Suvan J, Moles DR, Laurell L, Tonetti MS: Adjunctive benefits of systemic amoxicillin and metronidazole in non-surgical treatment of generalized aggressive periodontitis: a randomized placebo-controlled clinical trial. J Clin Periodontol 2005;32:1096–1107.
66 Matarazzo F, Figueiredo LC, Cruz SE, Faveri M, Feres M: Clinical and microbiological benefits of systemic metronidazole and amoxicillin in the treatment of smokers with chronic periodontitis: a randomized placebo-controlled study. J Clin Periodontol 2008;35:885–896.
67 Hallmon WW, Rees TD: Local anti-infective therapy: mechanical and physical approaches: a systematic review. Ann Periodontol 2003;8:99–114.
68 Lang NP, Tan WC, Krahenmann MA, Zwahlen M: A systematic review of the effects of full-mouth debridement with and without antiseptics in patients with chronic periodontitis. J Clin Periodontol 2008;35:8–21.
69 Sahrmann P, Puhan MA, Attin T, Schmidlin PR: Systematic review on the effect of rinsing with povidone-iodine during nonsurgical periodontal therapy. J Periodontal Res 2009;45:153–164.
70 Mombelli A, Schmid B, Rutar A, Lang NP: Persistence patterns of *Porphyromonas gingivalis*, *Prevotella intermedia/nigrescens*, and *Actinobacillus actinomycetemcomitans* after mechanical therapy of periodontal disease. J Periodontol 2000;71:14–21.
71 Magnusson I, Lindhe J, Yoneyama T, Liljenberg B: Recolonization of a subgingival microbiota following scaling in deep pockets. J Clin Periodontol 1984;11:193–207.
72 Sbordone L, Ramaglia L, Gulletta E, Iacono V: Recolonization of the subgingival microflora after scaling and root planing in human periodontitis. J Periodontol 1990;61:579–584.
73 Hanes PJ, Purvis JP: Local anti-infective therapy: pharmacological agents: a systematic review. Ann Periodontol 2003;8:79–98.
74 Bonito AJ, Lux L, Lohr KN: Impact of local adjuncts to scaling and root planing in periodontal disease therapy: a systematic review. J Periodontol 2005;76:1227–1236.
75 Heasman PA, Vernazza CR, Gaunt FL, Pennington MA: Cost-effectiveness of adjunctive antimicrobials in the treatment of periodontitis. Periodontol 2000 2011;55:217–230.
76 Kaner D, Bernimoulin JP, Hopfenmuller W, Kleber BM, Friedmann A: Controlled-delivery chlorhexidine chip versus amoxicillin/metronidazole as adjunctive antimicrobial therapy for generalized aggressive periodontitis: a randomized controlled clinical trial. J Clin Periodontol 2007;34:880–891.

77 Schwarz F, Aoki A, Becker J, Sculean A: Laser application in non-surgical periodontal therapy: a systematic review. J Clin Periodontol 2008;35:29–44.
78 Karlsson MR, Diogo Lofgren CI, Jansson HM: The effect of laser therapy as an adjunct to non-surgical periodontal treatment in subjects with chronic periodontitis: a systematic review. J Periodontol 2008;79: 2021–2028.
79 Azarpazhooh A, Shah PS, Tenenbaum HC, Goldberg MB: The effect of photodynamic therapy for periodontitis: a systematic review and meta-analysis. J Periodontol 2010;81:4–14.
80 Pfizer: Flagyl® 375 metronidazole capsule. 2006 (accessed 12 October 2009). www.pfizer.com/pfizer/download/uspi_flagyl_375.pdf.
81 GlaxoSmithKline: Prescribing information Amoxil® (amoxicillin). 2006.
82 Bendesky A, Menendez D, Ostrosky-Wegman P: Is metronidazole carcinogenic? Mutat Res 2002;511: 133–144.
83 Giannopoulou C, Andersen E, Brochut P, Plagnat D, Mombelli A: Enamel matrix derivative and systemic antibiotics as adjuncts to non-surgical periodontal treatment: biologic response. J Periodontol 2006;77: 707–713.
84 D'Aiuto F, Nibali L, Parkar M, Suvan J, Tonetti MS: Short-term effects of intensive periodontal therapy on serum inflammatory markers and cholesterol. J Dent Res 2005;84:269–273.
85 D'Aiuto F, Ready D, Tonetti MS: Periodontal disease and C-reactive protein-associated cardiovascular risk. J Periodontal Res 2004;39:236–241.
86 Noack B, Genco RJ, Trevisan M, Grossi S, Zambon JJ, De Nardin E: Periodontal infections contribute to elevated systemic C-reactive protein level. J Periodontol 2001;72:1221–1227.
87 Ardila CM, Granada MI, Guzman IC: Antibiotic resistance of subgingival species in chronic periodontitis patients. J Periodontal Res 2010;45:557–563.
88 Schrag SJ, Pena C, Fernandez J, Sanchez J, Gomez V, Perez E, Feris JM, Besser RE: Effect of short-course, high-dose amoxicillin therapy on resistant pneumococcal carriage: a randomized trial. JAMA 2001; 286:49–56.
89 Cunha-Cruz J, Hujoel PP, Maupome G, Saver B: Systemic antibiotics and tooth loss in periodontal disease. J Dent Res 2008;87:871–876.
90 Mombelli A, Cionca N, Almaghlouth A: Does adjunctive antimicrobial therapy reduce the perceived need for periodontal surgery? Periodontol 2000 2011;55:205–216.
91 Nishimura RA, Carabello BA, Faxon DP, Freed MD, Lytle BW, O'Gara PT, O'Rourke RA, Shah PM, Bonow RO, Chatterjee K, de Leon AC Jr, Gaasch WH, Otto CM, Shanewise JS, Smith SC Jr, Jacobs AK, Buller CE, Creager MA, Ettinger SM, Krumholz HM, Kushner FG, Page RL, Tarkington LG, Yancy CW Jr: ACC/AHA 2008 guideline update on valvular heart disease: focused update on infective endocarditis: a report of the American College of Cardiology/American Heart Association Task Force on Practice Guidelines: endorsed by the Society of Cardiovascular Anesthesiologists, Society for Cardiovascular Angiography and Interventions, and Society of Thoracic Surgeons. Circulation 2008; 118:887–896.
92 Berbari EF, Osmon DR, Carr A, Hanssen AD, Baddour LM, Greene D, Kupp LI, Baughan LW, Harmsen WS, Mandrekar JN, Therneau TM, Steckelberg JM, Virk A, Wilson WR: Dental procedures as risk factors for prosthetic hip or knee infection: a hospital-based prospective case-control study. Clin Infect Dis 2010;50:8–16.
93 Duval X, Alla F, Hoen B, Danielou F, Larrieu S, Delahaye F, Leport C, Briancon S: Estimated risk of endocarditis in adults with predisposing cardiac conditions undergoing dental procedures with or without antibiotic prophylaxis. Clin Infect Dis 2006; 42:e102–e107.
94 Esposito M, Grusovin MG, Talati M, Coulthard P, Oliver R, Worthington HV: Interventions for replacing missing teeth: antibiotics at dental implant placement to prevent complications. Cochrane Database Syst Rev 2008;3:CD004152.
95 Uckay I, Hoffmeyer P, Trampuz A, Borens O, Terzic A, Scolozzi P, Peter R: Antibiotic prophylaxis before dental procedures in arthroplasty patients (in German). Rev Med Suisse 2010;6:727–730.
96 Alekshun MN: New advances in antibiotic development and discovery. Expert Opin Investig Drugs 2005;14:117–134.
97 Eliopoulos GM: Microbiology of drugs for treating multiply drug-resistant gram-positive bacteria. J .Infect 2009;59(suppl 1):S17–S24.
98 Vooturi SK, Firestine SM: Synthetic membrane-targeted antibiotics. Curr Med Chem 2010;17:2292–2300.

Prof. Andrea Mombelli
School of Dental Medicine, University of Geneva
19 rue Barthélemy-Menn
CH–1205 Geneva (Switzerland)
Tel. +41 22 379 4030, E-Mail Andrea.Mombelli@unige.ch

Kinane DF, Mombelli A (eds): Periodontal Disease. Front Oral Biol. Basel, Karger, 2012, vol 15, pp 149–159

Regenerative Periodontal Therapy

Daniel W.K. Kao · Joseph P. Fiorellini

Department of Periodontics, School of Dental Medicine, University of Pennsylvania, Philadelphia, Pa., USA

Abstract

Traditional treatment for loss of bone and attachment due to periodontal disease has focused around repairing the damage induced. However, over the past few decades, clinicians have begun to utilize regenerative techniques to rebuild bone, cementum and the periodontal ligament. Conventional procedures most often involve the use of barrier membranes with bone grafts that foster selective cell repopulation and regrowth of osseous structures. Since the predictability of these techniques may be limited to certain case types, pharmacologically based efforts are underway to investigate the possibility of harnessing osseous regrowth potential. Clinical research has found that proteins are potent biological mediators that promote many of the events in wound healing, and have been shown to promote bone formation in human clinical studies.

Conventional periodontal therapy, including surgical and non-surgical approaches, may result in tissue shrinkage after healing [1]. In advanced defects, bone contouring may be applied during the surgical procedures in order to create a maintainable environment. This bone contouring approach may also create esthetic concerns, especially in the anterior area. However, treatments without performing bone contouring may result in residual pockets and create an inaccessible environment for proper cleaning procedures [2]. In addition, the risks associated with the healing process of conventional periodontal treatment can result in the growth of a long junctional epithelium attachment; this secondary structure may be a disadvantage due to a tendency to form new pockets [3, 4].

At the micro-structural level, the epithelium and its surrounding connective tissue often matures into various non-functional types of scar tissue after conventional periodontal therapy; this result is considered as a 'repair' part of the healing process [5]. These adverse effects of conventional periodontal therapy can be avoided or reduced by applying regenerative procedures to restore the loss of periodontal tissue [6]. The regenerative treatment procedures are designed to restore lost parts of the tooth-supporting structures. Several studies have demonstrated that regenerative periodontal procedures give more favorable results compared to conventional periodontal therapies [7–9].

In comparison to simple repair, periodontal regeneration is an ideal healing process that occurs through the reconstruction of new periodontal attachment apparatus, which involves the formation of new alveolar bone, a functionally aligned periodontal ligament (PDL) and new cementum [10]. Several studies, including meta-analysis results, have shown that periodontal regeneration procedures tend to show more clinical improvements than conventional procedures [11–15].

Regenerative periodontal procedures require biomaterials that lead to regeneration of the PDL, new attachments, new cementum and new alveolar bone. An ideal tissue engineering material should deliver the regenerative signals in an optimal manner to stimulate a cellular response. Initial efforts to achieve guided tissue regeneration (GTR) have aimed to regenerate attachment apparatus by isolating the defect with the barrier membranes and/or had grafting to maintain space so that regeneration could have enough time to take place. GTR is based on the exclusion of connective tissue and epithelium in favor of PDL regeneration, following the establishment of new attachments. Thus, GTR is the purposeful selection of cell types that repopulate at the wound with the intention of directing the healing tissue composition. Recently, the stimulation of periodontal regeneration with growth factors has become an effective and predictable technique. Based on the biological enhancement of wound healing, these molecules produce a true histological regeneration. Growth/regeneration factors and differentiation factors – such as enamel matrix proteins (amelogenin), growth factors (like platelet-derived growth factor; PDGF), as well as polypeptide mitogens (like bone morphogenic protein-2; BMP-2) – are being used to stimulate regeneration. These molecules augment and/or stimulate the natural healing response, and include stimulatory effects on angiogenesis, cellular differentiation, cellular proliferation and extracellular matrix biosynthesis, respectively. The stimulation by signaling molecules (growth factors) has improved predictability.

Periodontal Regenerative Techniques

Periodontal therapy has evolved from mainly resective concepts to regenerative treatments. The advantage of a regenerative approach not only includes a gain in clinical attachment, but the reduction or elimination of local factors. These include minimizing anatomical factors – such as root concavities, deep interproximal defects or furcations – where periodontal breakdown can easily progress. In addition, esthetic enhancements can be achieved, especially in the anterior regions.

Guided Tissue Regeneration

GTR includes multiple components, such as new attachment, reattachment, regeneration and repair. The term was first used when a new connective tissue attachment was demonstrated with human histology utilizing an expanded polytetrafluoroethylene (ePTFE) membrane (non-resorbable) [16]. 'New attachment' is the union of connective tissue or epithelium with the root surface which has been deprived of its original attachment. 'Reattachment' is the reunion of epithelium or connective tissue with root surfaces and bone, such as occurs after incision or injury. 'Repair' is the healing of a wound with periodontium that has not been fully restored to its original form and function. Lastly, 'true regeneration' is the histological term that describes the reconstruction of bone, cementum and the PDL.

The occlusive-based regeneration of the periodontium is primarily dependent on epithelial exclusion. Exclusion of epithelium is essential in barrier membrane regeneration. Barrier materials include natural absorbable polymers, such as collagen (types I, II, III, IV), collagen and GAG copolymer, synthetic absorbable polymers (such as polylactic acid and polyglycolic acid), fibrin, non-resorbable polymer PTFE and titanium-reinforced ePTFE. Generally, cementum deposition and PDL formation is slower than the downgrowth migration of epithelium. This natural process limits or prohibits regeneration and results in repair or a long junctional epithelium. In a novel investigation, Nyman et al. [6] demonstrated that the placement of a Millipore filter excluded the epithelium from the healing site. This exclusion with the addition of space creation/maintenance is essential in effective regeneration. Once epithelium exclusion is accomplished, wound healing originates from the PDL [17]. The peak mitotic activity of the PDL occurs at approximately 3 weeks [18]. The importance of the PDL cell response in GTR is also important to cementum formation. With cementum formation, the PDL fibers are also able to insert and attach. Therefore, cementum formation can be viewed as the rate-limiting step in periodontal regeneration [19].

The clinical outcome of GTR regeneration is dependent on the periodontal defect morphology, remaining periodontium and practitioner skill. Membrane GTR is a challenging procedure. The success of the procedure often is related to the premature exposure of the membrane to the oral cavity. Soft tissue management generally includes concepts that maximize volume, such as intrasulcular and releasing incisions. Following flap reflection, the surgical site must be fully debrided and the root surfaces prepared. The next surgical step is trimming and adapting the membrane. Osseous architecture may require conservative osteoplasty to accommodate a passively fitting membrane. For closure of the site, suture material such as ePTFE or polyglycolic acid should be chosen since these retain the least amount of plaque. Lastly, a periodontal dressing may be used to secure the surgical area, especially from food impaction or trauma. Postoperative care is important and patients are typically prescribed an antibiotic, anti-inflammatory and antibacterial rinse during the initial healing period.

Developmental clinical studies have demonstrated that GTR is a predictable procedure. Predictability in these studies was related to the severity and position of the furcation. Class II furcations and maxillary buccal furcations are more predictable

[20–22]. When graft material is added for space maintenance, a GTR and demineralized freeze-dried bone allograft combination produced more defect filling, pocket depth reduction and greater levels of clinical attachment [23]. Depending on the number of bone walls in the defect, GTR showed 95% filling of 3-wall defects, 82% filling of 2-wall defects, and 39% of 1-wall defects [24]. In addition, sites with the most disease showed the greatest benefit [25]. Long-term stability of periodontal regeneration has been evaluated, and the results showed that if the patient is under a professional supportive periodontal maintenance program and maintains good oral hygiene, the newly regenerated tissue can be maintained in the long term [2, 7, 26–30].

Signaling Molecules in Periodontal Tissue Regeneration

There has been mounting evidence that the material utilized as graft material may play a significant role in altering wound healing, from repair to regeneration. Analyses of autografts and allografts have found that signaling molecules contained in the material may be able to stimulate regeneration. The molecular basis of these factors has been extensively identified and characterized. As a result, several signaling molecules are currently being investigated in preclinical and clinical studies.

Enamel Matrix Proteins

Clinical investigations support periodontal regeneration with enamel matrix proteins or their derivative, enamel matrix derivative (EMD). The main constituent of this product is the amelogenin protein. Commercially available EMD is isolated from porcine tooth buds [31]. These porcine-derived proteins are very similar to proteins expressed during human enamelogenesis [32, 33]. Amelogenin has been shown to have a positive effect on acellular cementum, the PDL and alveolar bone development. The biological response of PDL cells to amelogenin in vitro demonstrates that RNA expression is stimulated, cellular functions are improved and metabolism is increased [34, 35]. In addition, enamel proteins have been shown to facilitate new blood vessel formation. EMD also has an effect on osseous formation/resorption. Osteoclastic activity and resorption are limited by EMD treatment, whereas osteoblastic activity is upregulated. Moreover, important osteoblastic activity markers were also enhanced at 2 and 3 weeks as an indication of continued biological activity [36]. A preclinical mouse model demonstrated new cellular cementum-like tissue formed along enamel-matrix-protein-treated root slices [37]. EMD treatment of human gingival fibroblasts has found a reduction in cell death from TNF and produces a tissue inhibitor of MMP-2 [38, 39].

The clinical application of EMD was compared to traditional GTR in lower molar class II buccal furcation defects. Results indicated an equivalent reduction in horizontal probing at 14 months after surgery [40]. Similar improvements were found in a comparison of contralateral randomized infrabony defects using EMD versus

GTR with a resorbable membrane. These improvements in periodontal parameters were maintained for up to 8 years [41]. Case reports have demonstrated significant periodontal regeneration with EMD [42]. One such report describes the treatment of a localized bone defect associated with a palatal groove [43]. In a 1-year follow up, the site maintained an 8-mm clinical attachment gain and a 2-mm residual probing depth. Esposito et al. [44] reported in a Cochrane systematic review of intrabony defects that EMD significantly improved attachment levels and reduced probing pocket depths when compared to placebo or controls. The authors also indicated that there was great heterogeneity in the trails, suggesting that the significant difference found should be interpreted with caution.

Platelet-Derived Growth Factor
PDGF, formed by the combination of two amino acid chains, are dimeric structural protein ligands of about 30,000 kDa in molecular weight. The nomenclature for their characterization uses an alphabetical suffix (A–D), based on the amino acid homology. The 3D structure of the protein leads to the formation of either homo- or heterodimers (PDGF-AA, AB, BB, CC, DD) [45]. These dimer molecules have two receptor-binding epitopes that function as active sites to induce the proliferation of preosteoblastic cells, chemotaxis and matrix apposition [46]. The A and B chains of PDGF, which have been more extensively studied, are both synthesized as precursor molecules that undergo proteolytic processing. These attach only to cells that have receptors to accommodate them. The PDGF receptor, based on the most recent literature reviews, has two general types: a and b receptors [47]. PDGF was originally isolated because of its ability to promote proliferation of mesenchymal cells [48]. It has since been recognized as initiating additional cellular responses, such as cell migration and survival [49].

Periodontal Utility of Platelet-Derived Growth Factor

The potential of recombinant human PDGF (rhPDGF) in tissue engineering has probably made it one of the most thoroughly studied growth factors in periodontics [50]. Originally PDGF and insulin-like growth factor I (IGF-I) were found in combination to synergistically upregulate DNA and protein synthesis in osteoblasts, and have demonstrated mitogenic and chemotactic effects on the PDL and alveolar bone cells [51–56]. Histological results have shown that PDGF/IGF-1 treated sites showed the formation of new attachment apparatus, while the control carrier gel group indicated the presence of long junctional epithelium without any clear regeneration [57]. In another study, rhPDGF-BB was used alone (without IGF), with bone allograft as the carrier, in patient teeth with interproximal intrabony defects and molar class II furcation defects and poor-to-hopeless prognosis requiring extraction. A notch was placed at the base of the defect on the root surfaces, osseous bony defects were then filled with demineralized freeze-dried bone allograft with 1 of 3 concentrations of rhPDGF-BB (0.5, 1.0 or 5.0 mg/ml). The hopeless teeth and surrounding tissue were taken en block after 9 months. Histomorphometric analyses were carried out in reference to the notch placed on the root surface. In the rhPDGF/allograft-treated defects, the vertical probing depth reductions as well as clinical attachment level gain for interproximal defects were approximately 6 mm, while radiographic bone fill was approximately 2 mm. Furcation defects treated with the rhPDGF/allograft demonstrated horizontal and vertical pocket depth reductions with a gain in clinical attachment levels of approximately 3 mm. The histological evaluation also indicated the regeneration of a complete new periodontal attachment apparatus. This study demonstrated that rhPDGF-BB alone can stimulate periodontal regeneration in both class II furcations and interproximal intrabony defects [51].

A prospective randomized clinical study to determine the safety and effectiveness of rhPDGF-BB was completed. rhPDGF-BB and β-tricalcium phosphate (β-TCP) were utilized for the treatment of advanced periodontal osseous defects and compared to the control. This multicenter study examined patients who required surgical treatment for a ≥4-mm intrabony periodontal defect; they were evaluated after 6 months. Experimental groups received β-TCP and either 0.3 or 1.0 mg/ml rhPDGF-BB, while the control group was given β-TCP alone. The clinical parameters used to assess PDGF-BB-induced effects included clinical attachment levels, gingival recession, radiographic linear bone growth and percent bone fill. Clinical attachment level gain was 3.8 mm for 0.3 mg/ml rhPDGF and significantly higher compared to β-TCP alone at 3 months following the surgical procedures, but at 6 months this difference was not statistically significant. The 6-month radiographic analysis showed linear bone gain of 2.6 mm and percent defect fill of 57% in the rhPDGF-BB-treated group (0.3 mg/ml), compared to 0.9 mm and 18%, respectively, in the control group [58]. The rhPDGF-BB-treated patients also demonstrated increased radiographic defect fill at 18–24 months after surgery compared to 6 months after surgery [59]. In summary, this multicenter study demonstrated the safety of rhPDGF-BB in the clinical treatment of periodontal defects.

The pharmacokinetics for growth factors used in periodontal regeneration have been optimized. The total load dose is dependent on protein stability, half-life and carrier properties. Gene therapy evaluation of PDGF has indicated that sustained release may amplify the intensity and duration of regeneration. In a small animal model, several different gene combinations for PDGF-A and B were utilized in critical size defects. Histomorphometric measurement found that PDGF-B stimulated cell proliferation compared to PDGF-A or control. Bone regeneration was greater for PDGF-B specimens compared to the other test and control groups. Gene expression was active in the periodontal defect for up to 3 weeks [60, 61]. PDGF-A was found to have a minimal effect on periodontal regeneration. Using similar gene therapy methods, cementoblasts were altered using PDGF-A gene application. These cells exhibited proliferation similar to continuous rhPDGF-AA protein application on native cementoblasts [62]. Gingival fibroblasts in 3D collagen lattices were evaluated for the effects of PDGF-A and PDGF-B gene transfer. Findings of this study indicated that cell repopulation and defect fill was stimulated by more than four times for the PDGF-B-gene-applied gingival fibroblasts, while PDGF-A and controls exhibited similarly diminished cellular activity [63].

Bone Morphogenetic Protein

Molecular regenerative therapy has also included the application of recombinant human BMP-2 (rhBMP-2). BMPs have been shown to be mediators of several processes in embryogenesis. The majority of BMPs are members of the transforming growth factor family. BMPs can function singly or together with other BMPs or cytokines. BMP-2 and 7 have been demonstrated to produce bone and periodontal regeneration. rhBMP-2 has become a standard in clinical practice for the regeneration of bone. Extensive preclinical investigations have provided a basis for this human indication. Goldhaber [64] found that a diffusible substance induced bone formation across a membrane. Urist et al. [65] implanted this substance in the soft tissue of a rodent and bone was formed in a non-osseous site. Later, BMP/TCP carrier was surgically implanted in critical size defects in dogs. rhBMP-2 stimulated sites demonstrated more than 90% new bone; the control group demonstrated new bone formation below 10% [65]. More recently, the human clinical application of rhBMP-2 has focused on extraction socket preservation and sinus augmentations. A randomized masked placebo-controlled multicenter clinical study demonstrated that the novel combination of rhBMP-2 and a commonly utilized absorbable collagen sponge (ACS) had a striking effect on de novo osseous formation for the placement of dental implants [66]. In addition, the efficacy of different rhBMP-2 concentrations in the anterior maxilla for bone regeneration was assessed with radiographic evaluation. CT evaluations found a significant difference in bone formation between subjects treated with a concentration of 1.5 mg/ml rhBMP-2 and the other groups [67]. In the lateral sinus augmentation procedure, 1.50 mg/ml rhBMP-2/ACS was compared to a standard autograft. Sinuses with an alveolar ridge height <6 mm were included in the study. A total of 160 patients were enrolled in the study. The results indicated that the mean changes in bone height were similar; however, the rhBMP-2/ACS-induced bone was significantly denser on histological evaluation [68].

The rhBMP-2 regeneration of the periodontal apparatus has been extensively investigated in preclinical models; however, no human investigations have been published. rhBMP-2 periodontal regeneration in rodents was evaluated for early wound healing at 10 and 38 days. New bone formation was significantly increased 10 days after rhBMP-2 application on experimental periodontal wounds, as confirmed with histological sections. These exhibited new bone, cementum and collagen fiber formation. Importantly, more cementum growth coronally was detected with rhBMP-2 application. Complete healing without any evidence of ankylosis occurred in all 38-day-treated sites [69]. Wikesjo et al. [70] found that in a canine model 0.02 mg/ml rhBMP-2 fostered periodontal regeneration. Bone height was significantly increased with the treatment as compared to the control (3.8 vs. 0.7 mm). However, there was limited cementum regeneration and ankylosis was noted in the rhBMP-2-treated sites. In a recent study, rhBMP-2/ACS was utilized in the treatment of 3-wall intrabony defects in the dog model. Surgical implantation of the rhBMP-2/ACS resulted in enhanced bone formation, but no improvement in cementum regeneration. As a result, there was minimal functional periodontal formation. The application of BMP gene vectors for bone and periodontal regeneration has been documented. Betz et al. [71] documented that a dose-dependent rhBMP-2 vector (adenovirus) for critical-size defects was on the femora of rats. The evaluations with radiographs and histology indicated that all of the defects treated with a high dose were bridged with bone, whereas in the medium and low dose groups, this occurred significantly less.

In conclusion, the goal of these various treatment modalities is to regenerate periodontal tissues lost secondary to the disease. Current regenerative periodontal therapies utilize the surgical placement of biomaterials and cytokines. The transitions from barrier membranes to growth-factor-mediated regeneration enhance predictably, are easy to use, and reduce complications.

References

1 Isidor F, Karring T, Attstrom R: The effect of root planing as compared to that of surgical treatment. J Clin Periodontol 1984;11:669.

2 Lindhe J, Lang NP, Karring T: Clinical Periodontology and Implant Dentistry, ed 5. Oxford, Blackwell, 2008.

3 Rose LF: Periodontics: Medicine, Surgery, and Implants. St. Louis, Mosby, 2004.

4 Barrington EP: An overview of periodontal surgical procedures. J Periodontol 1981;52:518.

5 American Academy of Periodontology: Glossary of Periodontal Terms, ed 3. Chicago, The Academy, 1992.

6 Nyman S, Lindhe J, Karring T, Rylander H: New attachment following surgical treatment of human periodontal disease. J Clin Periodontol 1982;9:290.

7 Eickholz P, Hausmann E: Evidence for healing of interproximal intrabony defects after conventional and regenerative therapy: digital radiography and clinical measurements. J Periodontal Res 1998;33: 156.

8 Tonetti MS, Cortellini P, Suvan JE, et al: Generalizability of the added benefits of guided tissue regeneration in the treatment of deep intrabony defects: evaluation in a multi-center randomized controlled clinical trial. J Periodontol 1998;69:1183.

9 Eickholz P, Krigar DM, Kim TS, et al: Stability of clinical and radiographic results after guided tissue regeneration in infrabony defects. J Periodontol 2007;78:37.

10 American Academy of Periodontology: Glossary of Periodontal Terms. 2001. www.perio.org.

11 Murphy KG, Gunsolley JC: Guided tissue regeneration for the treatment of periodontal intrabony and furcation defects: a systematic review. Ann Periodontol 2003;8:266.

12 Needleman IG, Worthington HV, Giedrys-Leeper E, Tucker RJ: Guided tissue regeneration for periodontal infra-bony defects. Cochrane Database Syst Rev 2006:CD001724.

13 Laurell L, Gottlow J, Zybutz M, Persson R: Treatment of intrabony defects by different surgical procedures: a literature review. J Periodontol 1998;69: 303.

14 Camelo M, Nevins ML, Schenk RK, et al: Clinical, radiographic, and histologic evaluation of human periodontal defects treated with Bio-Oss and Bio-Gide. Int J Periodontics Restorative Dent 1998;18: 321.
15 Sculean A, Stavropoulos A, Windisch P, et al: Healing of human intrabony defects following regenerative periodontal therapy with a bovine-derived xenograft and guided tissue regeneration. Clin Oral Investig 2004;8:70.
16 Gottlow J, Nyman S, Lindhe J, et al: New attachment formation in the human periodontium by guided tissue regeneration: case reports. J Clin Periodontol 1986;13:604.
17 Aukhil I, Pettersson E, Suggs C: Guided tissue regeneration: an experimental procedure in beagle dogs. J Periodontol 1986;57:727.
18 Nyman S, Gottlow J, Karring T, Lindhe J: The regenerative potential of the periodontal ligament: an experimental study in the monkey. J Clin Periodontol 1982;9:257.
19 Caton JG, DeFuria EL, Polson AM, Nyman S: Periodontal regeneration via selective cell repopulation. J Periodontol 1987;58:546.
20 Pontoriero R, Celenza F Jr, Ricci G, Carnevale G: Rapid extrusion with fiber resection: a combined orthodontic-periodontic treatment modality. Int J Periodontics Restorative Dent 1987;7:30.
21 Pontoriero R, Lindhe J: Guided tissue regeneration in the treatment of degree II furcations in maxillary molars. J Clin Periodontol 1995;22:756.
22 Pontoriero R, Wennstrom J, Lindhe J: The use of barrier membranes and enamel matrix proteins in the treatment of angular bone defects: a prospective controlled clinical study. J Clin Periodontol 1999;26: 833.
23 Anderegg CR, Martin SJ, Gray JL, et al: Clinical evaluation of the use of decalcified freeze-dried bone allograft with guided tissue regeneration in the treatment of molar furcation invasions. J Periodontol 1991;62:264.
24 Cortellini P, Pini Prato G, Tonetti MS: Periodontal regeneration of human infrabony defects. I. Clinical measures. J Periodontol 1993;64:254.
25 Selvig KA, Kersten BG, Wikesjo UM: Surgical treatment of intrabony periodontal defects using expanded polytetrafluoroethylene barrier membranes: influence of defect configuration on healing response. J Periodontol 1993;64:730.
26 Gottlow J, Nyman S, Karring T: Maintenance of new attachment gained through guided tissue regeneration. J Clin Periodontol 1992;19:315.
27 McClain PK, Schallhorn RG: Long-term assessment of combined osseous composite grafting, root conditioning, and guided tissue regeneration. Int J Periodontics Restorative Dent 1993;13:9.
28 Cortellini P, Pini-Prato G, Tonetti M: Periodontal regeneration of human infrabony defects (V): effect of oral hygiene on long-term stability. J Clin Periodontol 1994;21:606.
29 Cortellini P, Paolo G, Prato P, Tonetti MS: Long-term stability of clinical attachment following guided tissue regeneration and conventional therapy. J Clin Periodontol 1996;23:106.
30 Machtei EE, Grossi SG, Dunford R, et al: Long-term stability of class II furcation defects treated with barrier membranes. J Periodontol 1996;67:523.
31 Hammarstrom L: Enamel matrix, cementum development and regeneration. J Clin Periodontol 1997; 24:658.
32 Gestrelius S, Andersson C, Lidstrom D, et al: In vitro studies on periodontal ligament cells and enamel matrix derivative. J Clin Periodontol 1997; 24:685.
33 Gestrelius S, Andersson C, Johansson AC, et al: Formulation of enamel matrix derivative for surface coating: kinetics and cell colonization. J Clin Periodontol 1997;24:678.
34 Barkana I, Alexopoulou E, Ziv S, et al: Gene profile in periodontal ligament cells and clones with enamel matrix proteins derivative. J Clin Periodontol 2007; 34:599.
35 Schlueter SR, Carnes DL, Cochran DL: In vitro effects of enamel matrix derivative on microvascular cells. J Periodontol 2007;78:141.
36 Galli C, Macaluso GM, Guizzardi S, et al: Osteoprotegerin and receptor activator of nuclear factor-kappa B ligand modulation by enamel matrix derivative in human alveolar osteoblasts. J Periodontol 2006;77:1223.
37 Song AM, Shu R, Xie YF, et al: A study of enamel matrix proteins on differentiation of porcine bone marrow stromal cells into cementoblasts. Cell Prolif 2007;40:381.
38 Zeldich E, Koren R, Dard M, et al: Enamel matrix derivative protects human gingival fibroblasts from TNF-induced apoptosis by inhibiting caspase activation. J Cell Physiol 2007;213:750.
39 Zeldich E, Koren R, Dard M, et al: Enamel matrix derivative induces the expression of tissue inhibitor of matrix metalloproteinase-3 in human gingival fibroblasts via extracellular signal-regulated kinase. J Periodontal Res 2010;45:200.

40 Hoffmann T, Richter S, Meyle J, et al: A randomized clinical multicentre trial comparing enamel matrix derivative and membrane treatment of buccal class II furcation involvement in mandibular molars. Part III: patient factors and treatment outcome. J Clin Periodontol 2006;33:575.
41 Sculean A, Schwarz F, Miliauskaite A, et al: Treatment of intrabony defects with an enamel matrix protein derivative or bioabsorbable membrane: an 8-year follow-up split-mouth study. J Periodontol 2006;77:1879.
42 Cortellini P, Tonetti MS: Minimally invasive surgical technique and enamel matrix derivative in intrabony defects. I: Clinical outcomes and morbidity. J Clin Periodontol 2007;34:1082.
43 Zucchelli G, Mele M, Checchi L: The papilla amplification flap for the treatment of a localized periodontal defect associated with a palatal groove. J Periodontol 2006;77:1788.
44 Esposito M, Grusovin MG, Papanikolaou N, et al: Enamel matrix derivative (Emdogain) for periodontal tissue regeneration in intrabony defects: a Cochrane systematic review. Eur J Oral Implantol 2009;2:247.
45 Reigstad LJ, Varhaug JE, Lillehaug JR: Structural and functional specificities of PDGF-C and PDGF-D, the novel members of the platelet-derived growth factors family. FEBS J 2005;272:5723.
46 Fiedler J, Roderer G, Gunther KP, Brenner RE: BMP-2, BMP-4, and PDGF-bb stimulate chemotactic migration of primary human mesenchymal progenitor cells. J Cell Biochem 2002;87:305.
47 Heldin CH, Ostman A, Ronnstrand L: Signal transduction via platelet-derived growth factor receptors. Biochim Biophys Acta 1998;1378:F79.
48 Heldin CH, Wasteson A, Westermark B: Platelet-derived growth factor. Mol Cell Endocrinol 1985; 39:169.
49 Claesson-Welsh L: Platelet-derived growth factor receptor signals. J Biol Chem 1994;269:32023.
50 Lynch SE, Williams RC, Polson AM, et al: A combination of platelet-derived and insulin-like growth factors enhances periodontal regeneration. J Clin Periodontol 1989;16:545.
51 Nevins M, Camelo M, Nevins ML, et al: Periodontal regeneration in humans using recombinant human platelet-derived growth factor-BB (rhPDGF-BB) and allogenic bone. J Periodontol 2003;74:1282.
52 Camelo M, Nevins ML, Schenk RK, et al: Periodontal regeneration in human class II furcations using purified recombinant human platelet-derived growth factor-BB (rhPDGF-BB) with bone allograft. Int J Periodontics Restorative Dent 2003;23:213.
53 Howell TH, Fiorellini JP, Paquette DW, et al: A phase I/II clinical trial to evaluate a combination of recombinant human platelet-derived growth factor-BB and recombinant human insulin-like growth factor-I in patients with periodontal disease. J Periodontol 1997;68:1186.
54 Wang HL, Pappert TD, Castelli WA, et al: The effect of platelet-derived growth factor on the cellular response of the periodontium: an autoradiographic study on dogs. J Periodontol 1994;65:429.
55 Taba M Jr, Jin Q, Sugai JV, Giannobile WV: Current concepts in periodontal bioengineering. Orthod Craniofac Res 2005;8:292.
56 Anusaksathien O, Jin Q, Zhao M, et al: Effect of sustained gene delivery of platelet-derived growth factor or its antagonist (PDGF-1308) on tissue-engineered cementum. J Periodontol 2004;75:429.
57 Lynch SE, de Castilla GR, Williams RC, et al: The effects of short-term application of a combination of platelet-derived and insulin-like growth factors on periodontal wound healing. J Periodontol 1991; 62:458.
58 Nevins M, Giannobile WV, McGuire MK, et al: Platelet-derived growth factor stimulates bone fill and rate of attachment level gain: results of a large multicenter randomized controlled trial. J Periodontol 2005;76:2205.
59 McGuire MK, Kao RT, Nevins M, Lynch SE: rhPDGF-BB promotes healing of periodontal defects: 24-month clinical and radiographic observations. Int J Periodontics Restorative Dent 2006;26:223.
60 Zhu Z, Lee CS, Tejeda KM, Giannobile WV: Gene transfer and expression of platelet-derived growth factors modulate periodontal cellular activity. J Dent Res 2001;80:892.
61 Jin Q, Anusaksathien O, Webb SA, et al: Engineering of tooth-supporting structures by delivery of PDGF gene therapy vectors. Mol Ther 2004;9:519.
62 Giannobile WV, Lee CS, Tomala MP, et al: Platelet-derived growth factor (PDGF) gene delivery for application in periodontal tissue engineering. J Periodontol 2001;72:815.
63 Anusaksathien O, Webb SA, Jin QM, Giannobile WV: Platelet-derived growth factor gene delivery stimulates ex vivo gingival repair. Tissue Eng 2003; 9:745.
64 Goldhaber P: Osteogenic induction across Millipore filters in vivo. Science 1961;133:2065.
65 Urist MR, Nilsson O, Rasmussen J, et al: Bone regeneration under the influence of a bone morphogenetic protein (BMP) beta tricalcium phosphate (TCP) composite in skull trephine defects in dogs. Clin Orthop Relat Res 1987;214:295.

66 Fiorellini JP, Howell TH, Cochran D, et al: Randomized study evaluating recombinant human bone morphogenetic protein-2 for extraction socket augmentation. J Periodontol 2005;76:605.
67 Bianchi J, Fiorellini JP, Howell TH, et al: Measuring the efficacy of rhBMP-2 to regenerate bone: a radiographic study using a commercially available software program. Int J Periodontics Restorative Dent 2004;24:579.
68 Triplett RG, Nevins M, Marx RE, et al: Pivotal, randomized, parallel evaluation of recombinant human bone morphogenetic protein-2/absorbable collagen sponge and autogenous bone graft for maxillary sinus floor augmentation. J Oral Maxillofac Surg 2009;67:1947.
69 King GN, King N, Cruchley AT, et al: Recombinant human bone morphogenetic protein-2 promotes wound healing in rat periodontal fenestration defects. J Dent Res 1997;76:1460.
70 Wikesjo UM, Xiropaidis AV, Thomson RC, et al: Periodontal repair in dogs: rhBMP-2 significantly enhances bone formation under provisions for guided tissue regeneration. J Clin Periodontol 2003;30:705.
71 Betz OB, Betz VM, Nazarian A, et al: Delayed administration of adenoviral BMP-2 vector improves the formation of bone in osseous defects. Gene Ther 2007;14:1039.

Prof. Joseph P. Fiorellini DMD, DMSc
Department of Periodontics, School of Dental Medicine, University of Pennsylvania
240 South 40th Street
Philadelphia, PA 19104 (USA)
E-Mail jpf@exchange.upenn.edu

Kinane DF, Mombelli A (eds): Periodontal Disease. Front Oral Biol. Basel, Karger, 2012, vol 15, pp 160–176

Paradigm Shift in the Pharmacological Management of Periodontal Diseases

Hatice Hasturk · Alpdogan Kantarci · Thomas E. Van Dyke

Department of Periodontology, The Forsyth Institute, Cambridge, Mass., USA

Abstract

It is becoming clear that variations in inflammatory response are a major determinant in susceptibility to periodontitis. However, our understanding of the relationship of the causal agents in periodontitis to the pathogenesis is not as clear as we once thought, and thus therapies based on etiopathogenesis are similarly in question. We are entering a new era of therapeutic discovery that may have a major impact on our management of the periodontal diseases. Fundamentally, periodontitis is an irreversible condition and once both soft and hard tissues are lost, the healthy periodontal architecture cannot be completely or predictably rebuilt. The discovery of new families of lipid mediators of resolution of inflammation (the lipoxins) and eicosapentaenoic-acid- and docosahexaenoic-acid-derived chemical mediators (the resolvins and protectins) opens new avenues to designing resolution-targeted therapies to control the unwanted side effects of excessive inflammation. The novel protective and therapeutic actions of pro-resolution lipid mediators following microbial challenge are mediated by regulation of the local and systemic inflammatory response that has a direct impact on the organization of the biofilm (plaque) and suggests a new paradigm in clinical periodontal therapeutics.

Periodontal infection is among the most common infections in man, and is considered to be initiated by microorganisms of the dental plaque biofilm [1, 2]. The pathogenesis of periodontal disease has for convenience often been described in terms of an infectious disease that one would 'catch' with similarities to other acquired exogenous infections [3, 4]. However, there is currently an inadequate understanding of host-related factors affecting the initiation and progression, and of the impact of environmental determinants of the disease, such that therapeutic interventions have focused on methods to eliminate the 'pathogenic' bacteria that were assumed by association to be directly involved in the destruction of the host tissues [5]. Hence, periodontal therapy has focused on the mechanical removal of plaque and its bacterial contents supplemented with a wide range of antibiotic and antiseptic compounds [6, 7]. Mechanical debridement, a widely-used and accepted technique to treat and prevent the future incidence of periodontitis, has been regarded as the 'gold standard' treatment, and it is generally accepted that chemical antibacterial treatments alone do not have efficacy in periodontal treatment [8]. Mechanical methods, including nonsurgical and surgical techniques, aim to eliminate bacterial biofilms and prevent regrowth by removing ecological niches [9]. Practitioners of dentistry, and particularly periodontists, have come to appreciate that permanent eradication of bacteria is not possible [10, 11]. Moreover, with the development of more sensitive techniques for the identification of bacteria [12], it has been hypothesized that the pathogenic biofilm in periodontitis represents an overgrowth of commensal organisms [13]. The parameters that enable the shift in the composition of the biofilm highlight the multi-faceted complexity of periodontal pathogenesis [14]. These insights do not reduce the importance of bacteria and other microorganisms in the etiology and pathogenesis of periodontal disease; however, host-related factors – such as genetics and immune response, diet, smoking, stress, systemic health and social elements – are all known to be major disease determinants [15, 16]. Thus, a fresh and unbiased assessment is required to design novel treatment modalities in complex inflammatory diseases such as periodontitis. This chapter focuses on the shift in emphasis from the periodontal infection to periodontal inflammation as the target for pharmacological interventions for treatment of periodontal diseases.

What Is Unique about Periodontal Diseases?

Anatomically, teeth are the only mineralized structures that traverse the epithelium to be exposed to an extremely contaminated external environment [17]. The soft and hard tissues of the periodontium include various types of epithelium, connective tissue, bone and a resilient attachment of the teeth to these structures (the periodontal ligament). Within this sophisticated organ system, the tooth serves not only as a medium for external stimuli, but its surface is also an environment for accumulation and colonization by a diverse group of bacterial species. Because teeth are always exposed to the external environment, the periodontium is continuously contaminated by bacteria [1]. The interaction between dental-periodontal tissues and colonizing organisms is a lifelong process that begins immediately after tooth eruption. However, considering this exposure, there is relatively little disease; a balance between host and bacteria is generally maintained in this commensal relationship. Periodontal diseases are chronic diseases that result from host-mediated destruction of the supporting tissues of the dentition when this balance is lost or disturbed by genetic and environmental risk factors [18, 19].

Evolution of the Current Treatment Paradigm for Periodontal Diseases

A link between bacteria and disease was discovered as early as the 18th and 19th centuries (germ theory) [20]. This was a major breakthrough in the history of medicine.

Koch described the direct link between certain microorganisms and disease by demonstrating that once the associated bacteria were absent, the disease resolved [21]. This observation combined with the discovery of penicillin in the 20th century formed the basis of therapy, and successfully controlled or eliminated major infectious diseases [22]. However, the assumption that there was an antibiotic for every infection was an oversimplification of the complexity of the microbiome and interspecies ecological dynamics of microorganisms. It was soon realized that not all bacterially associated diseases were treatable with antibiotics. The human microbiome comprises more bacterial cells than host cells and the symbiotic balance is central to health. The microbiome is not static; the species change and adapt to environmental challenges, most notably, the emergence of antibiotic resistance [23]. We continue to define the human microbiome, and we continue to search for effective methods to restore homeostasis between the host and commensal bacteria.

The concept of removal of accretions on the teeth is much older than the germ theory. Elimination of the accumulation of debris on tooth surfaces was associated with good health was recognized by the ancients; Hippocrates recommended that 'teeth be cleaned with a ball of wool'. While the mechanism of why this was beneficial was not immediately clear to our ancestors, even the first practicing dentists used this approach in the routine prevention and treatment of pyorrhea [24]. Various instruments were developed over time for use by dental practitioners in a professional setting and for patient home use to remove what we now call 'plaque' build-up on tooth surfaces [25]. After thousands of years, this is still the most essential component of oral hygiene measures for the prevention of periodontal diseases.

The demonstration that bacteria were causative in periodontal disease was not made until 1960 [26]. Subsequently, the discovery of the association of specific bacterial species with health or disease and detailed characterization of various complexes of bacteria led to current antimicrobial approaches for prevention and treatment of periodontitis. In his pioneering work, Socransky [27] refined Koch's postulates in the context of periodontitis and established our modern understanding of periodontal infection. Both non-surgical and surgical periodontal therapies have the goal of elimination of periodontal pockets to provide a cleansable environment for the maintenance of a plaque-free dentition [28]. The variety of periodontal therapeutic procedures is beyond the scope of this chapter; the important point is that the bacteria are the target. Various antiseptic, systemic and local antibiotic regimens have been added in conjunction with mechanical techniques to further target the bacteria with considerable success [29]. Nevertheless, periodontitis continues to present a significant health burden in the susceptible population [30].

Not too long ago, it was assumed that periodontitis could be eliminated in modern societies though better methods of plaque control; with time, only under-served populations in the developing world would have significant periodontal disease. Interestingly, in recent years, it has been documented that diseases of aging, including periodontitis, are negatively impacted by increasing obesity, smoking and other behaviors of modern society, and that there are interactions between diseases of aging, including type 2 diabetes, cardiovascular diseases and periodontitis [31–33]. The association of periodontitis with major non-infectious diseases suggests that a better understanding of the host's role in periodontal susceptibility is necessary.

Role of Inflammation in Periodontal Diseases

Inflammation is a physiologic series of responses generated by the host in response to infection or injury. The initial events of inflammation are initiated by vascular reactions at the site of injury leading to extravasation of fluid and plasma proteins and recruitment of leukocytes to the site of injury [34]. Inflammation is a natural beneficial response confining the injury or infection, and serves as the first step in the initiation of the immune response [35]. Ideally, the innate immune or inflammatory response is sufficient to eliminate the infection, allowing for healing and return to homeostasis. If the innate immune response fails, chronic inflammation ensues. In the case of chronic inflammation, the persistence of the response leads to host tissue self-destruction and may result in irreversible pathological changes. Various cell types, including mast cells, platelets and leukocytes generate inflammatory mediators (chemokines and cytokines). Elevated levels of these molecules promote inflammation and amplify the response [35]. Proteins – including cytokines and chemokines, low-molecular-weight lipids derived from arachidonic acid, nitric oxide and carbon monoxide, reactive oxygen species, and nucleotides – are recognized as inflammatory mediators [36, 37].

Neutrophils are the most abundant white blood cells recruited to the early inflammatory periodontal lesion [18, 38]. The early response that initiates the inflammatory cascade begins with expression of cell adhesion molecules on endothelial cells and secretion of various cytokines such as IL-8 and MCP-1 in response to the bacterial stimulus that recruits neutrophils to the site of invasion [39, 40]. If the first line of defense fails to clear the bacteria, the lesion matures as described histologically by Page and Schroeder [41]: the established chronic gingivitis lesion exhibiting all the components of an immune lesion with T cells, B cells and antibody-secreting plasma cells. The trigger for transition from the established (stable) gingivitis lesion to the advanced or progressing periodontitis lesion remains elusive, but it is known that it is characterized by a continuous and uncontrolled response by neutrophils. Interestingly, it would appear that in the progressing immune lesion of periodontitis, there is continuous neutrophil activation leading to 'neutrophil-mediated tissue injury' [15, 42, 43]. Localized aggressive periodontitis (fig. 1a, b) has been called the 'Rosetta Stone' of periodontal diseases providing insights into the pathogenesis of more common forms of chronic periodontitis [38]. The localized aggressive periodontitis lesion is an example of neutrophil-mediated tissue injury in which neutrophils are primed or hyperactivated resulting in elevated basal levels of superoxide generation and an increase in response to secondary stimuli (fig. 1c, d). In chronic periodontitis, the

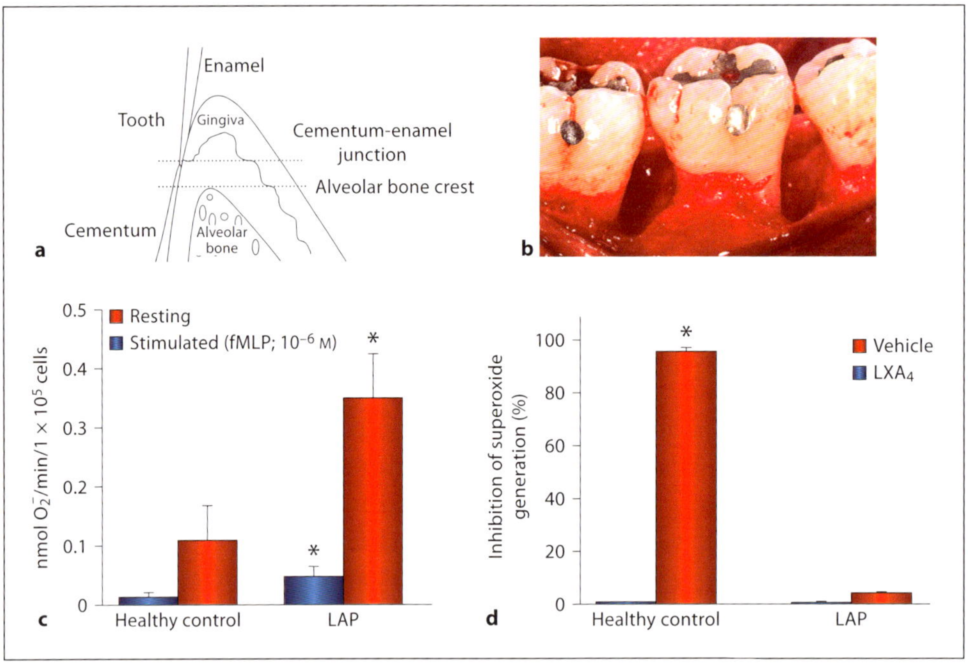

Fig. 1. Localized aggressive periodontitis (LAP) is characterized by severe and localized bone loss around first molar and incisor teeth. **a** Schematic depiction of periodontal structures in LAP. **b** Clinical appearance of bone around molar teeth after the gingival flaps are elevated during periodontal surgery clearly shows the extent of localized destruction. **c** LAP neutrophils present a primed phenotype as shown by elevated basal levels of superoxide generation that lead to further excessive response to secondary challenge. * $p < 0.05$. **d** In response to LXA_4 treatment, agonist-induced superoxide generation is significantly inhibited in neutrophils from non-LAP healthy individuals that were chosen as matched controls, while LAP neutrophils do not respond to LXA_4 suggesting a failure of resolution pathways. * $p < 0.05$.

trigger for conversion of the gingivitis lesion to periodontitis is likely related to excessive recruitment of neutrophils followed by excessive release of inflammatory mediators associated with rapid and widespread tissue destruction [42].

Once the conversion of the gingivitis lesion occurs, the neutrophil is a rich source of prostaglandin (PG)E_2 that probably accounts for the majority of PGE_2 detected in the gingival crevicular fluid of periodontitis patients [44]. Levels of PGE_2 in the crevicular fluid of patients with periodontal infections correlate with disease severity and aggressiveness [45].

Lipid Mediators of Inflammation

Initially isolated from peanut *(Arachis)* oil, arachidonic acid has been shown to be a central component of phospholipid metabolism in mammals. It is generated from linoleic acid by the action of a desaturase enzyme system [46, 47]. Biosynthesis of most eicosanoids begins with the cleavage of arachidonic acid from the 2-position of phospholipids in the cell membrane through the action of phospholipase A_2 (PLA_2). Prostanoids and leukotrienes (LT) are produced via the action of the enzymes cyclooxygenase (COX) and lipoxygenase (LO) [49]. COX has two isoenzymes: COX-1 and COX-2, sharing 64% identity and similar enzymatic activity [48]. While COX-1 is expressed constitutively, COX-2 is activated by ligand-coupled stimulation and alternatively known as the inducible or inflammatory COX [49]. The second major group of eicosanoids comprise the LT and other products of the LO pathways [50]. Three different LO substitute the 5, 12 and 15 positions of arachidonic acid, giving rise to hydroperoxides of eicosatetraenoic acid (HPETE). The 5-LO, 12-LO and 15-LO are found in neutrophils, platelets and endothelial/epithelial cells, respectively, and their products are named accordingly (5-HPETE, 12-HPETE and 15-HPETE) [51–54].

Resolution of Inflammation

Since many molecules in eicosanoid metabolism are associated with proinflammatory outcomes, blocking the actions of arachidonic acid cascade products has been the major pharmacologic target for blocking inflammation. Many drugs have been developed to arrest or modify inflammation by inhibiting the enzymatic pathways or blocking the receptors that lead to the generation of lipid mediators. One action of corticosteroids is thought to be inhibition of PLA_2 blocking the release of arachidonic acid from membrane phospholipids [55]. This nonspecific inhibition of eicosanoid generation, although highly effective, is not without side effects leading to multiple adverse outcomes including suppression of the acquired immune system. Nonsteroidal anti-inflammatory drugs or specific COX-2 inhibitors have been used to block PG production [56–60] specifically by suppressing COX-mediated inflammation without blocking acquired immunity [56]. However, these agents also lead to significant side effects, due to their lack of selectivity, blocking both the constitutive (COX-1) and the inducible (COX-2) isoforms. There is considerable controversy in the clinical literature about the unwanted side effects of COX-2 inhibitors in the gastrointestinal and cardiovascular systems [61–63].

A major discovery by Serhan et al. [64] changed the prevailing concepts relating to the resolution of inflammation. Since the first descriptions of inflammation in ancient times to the late 20th century, it was always assumed that the inflammation passively resolves following the degradation or loss of proinflammatory mediators. Recent work shows that resolution of inflammation is an active process mediated by a newly identified class of eicosanoids produced by the actions of two different LO enzymes on a single arachidonic acid molecule. The product of this LO:LO interaction is a new class of molecule called lipoxins (LX) [65]. Several different pathways can generate LX. In general, interactions between cells result in the generation of LX, while single cells can also produce these compounds [66]. For example, arachidonic acid is substituted at the 15 carbon by endothelial cells and the resulting 15-HPETE is further acted upon by neutrophil 5-LO; the 5,15-HPETE is rapidly converted by hydrolases to either LXA_4 and/or LXB_4 [67–69]. LX are receptor agonists whose major functions include the arrest of neutrophil activity (migration and secretion), induction of neutrophil apoptosis, and the non-phlogistic recruitment of monocytes and macrophages. Non-phlogistic, or literally, non-fever-producing macrophages secrete no proinflammatory cytokine, but exhibit enhanced phagocytosis of bacteria and apoptotic neutrophils [37, 64, 70–73].

The neutrophil's role in LX generation is crucial. When stimulated by various agonists, neutrophils release 15-HETE, which is further transformed into LX. This pathway suggests that precursors of LX synthesis can be stored within the membranes of inflammatory cells and then released by stimuli activating PLA_2. The second major route of LX biosynthesis is mediated by the interaction of human neutrophils with platelets in blood vessels. Cell-to-cell interaction involves 5-LO in neutrophils and 12-LO in platelets for the insertion of molecular oxygen into arachidonic acid [74]. When platelets adhere to neutrophils, they convert LTA_4 into a cation (carbonium) by 12-LO [75] and produce LXB_4 or LXA_4 functioning as a 'LX synthase' in platelets [75]. Platelet-neutrophil adherence and transcellular biosynthesis are important anti-inflammatory events that regulate neutrophil recruitment by initiating the formation of lipid mediators that limit proinflammatory responses.

Aspirin also plays an important role in the generation of LX [72]. COX-2 inhibition by aspirin is by acetylation of COX-2, which generates a new still-active enzyme, 15R-LO. The enzymatic product is 15-*R*-HETE instead of PG, which forms the epimer15-R-LX (aspirin-triggered LX, ATLa). ATLa is more potent than native LXA_4 in inhibiting neutrophil adhesion and cell proliferation [76]. COX-2 is increased in inflammatory reactions and in disease states, including colon cancer, and normalized upon aspirin intake [77–79]. These actions are linked to ATLa that serve as endogenous anti-inflammatory signals that mediate some of aspirin's beneficial actions [77]. Aspirin also impacts COX-2-bearing vascular endothelial cells or epithelial cells and their coactivation with neutrophils [72].

Potential Role for Lipoxins as Therapeutics

While it is well appreciated today that neutrophils play an essential role in host defense, their prolonged activation can lead to tissue injury observed in common chronic inflammatory diseases. Because of the importance of leukocyte trafficking, endogenous positive and negative signaling mediators of inflammation have been referred to as local 'go' and 'stop' signals [35, 66, 80–84]. LXs are an important stop signals [85] produced in vivo during the course of inflammation [86–89]. Alterations in LX levels may be linked to the pathophysiology of several human diseases and may

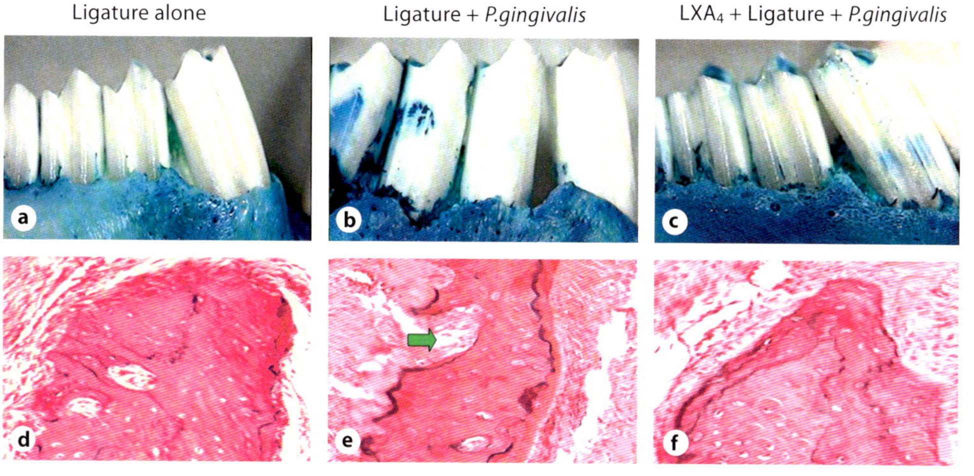

Fig. 2. LXA_4 prevents periodontal inflammation and bone loss induced by *P. gingivalis* in the rabbit periodontitis model. **a** As opposed to other animal models of periodontitis, ligature placement alone does not cause any change in alveolar bone. **b** In response to experimental periodontitis induced by *P. gingivalis* and ligature placement, local severe bone loss is observed. **c** Topical LXA_4 treatment simultaneously applied with *P. gingivalis* prevents periodontitis and alveolar bone loss. **d** Histological analysis on HE-stained sections reveals that LXA_4 prevents bone loss. **e** Periodontal disease induced by *P. gingivalis* results in alveolar bone resorption (depicted by arrow). **f** 15-LO transgenic rabbits are resistant to experimental periodontitis.

have specific local functions. The first demonstration of LX in clinical inflammation was in bronchial lavage. LXA_4 and LXB_4 are formed in nasal polyps and LXA_4 is generated in nasal lavage from aspirin-sensitive asthmatics and in experimental nephritis [90]. Rupture of atherosclerotic plaque leads to rapid generation of LXA_4 in the intracoronary artery [91]. LXs are also generated by normal human bone marrow, and during chronic myelocytic leukemia platelets lose 12-LO, losing the ability to generate LX, and this may be related to blast crisis [92]. In humans, the aspirin-tolerant subjects generated both LXA_4 and ATL, but aspirin-intolerant patients proved to have a diminished capacity to generate ATL and LX upon aspirin challenge [93]. In an experimental periodontitis rabbit model, animals are protected by LXA_4 and transgenic rabbits over-expressing 15-LO generate enhanced levels of LX, exhibit a reduced inflammatory phenotype, and are protected from bone loss in periodontal disease [94] (fig. 2).

Resolvins as Pharmacological Agents in Therapeutic Intervention of Periodontal Diseases

Dietary fatty acids, particularly the ω–3 polyunsaturated fatty acids (ω–3 PUFA), are known to clinically reduce inflammation in humans. Mammals do not produce the major ω–3 fatty acids found in fish oil: eicosapentaenoic acid (EPA) and its derivative docosahexaenoic acid (DHA). EPA and DHA are held to be beneficial in a wide range of human inflammatory disorders, including cardiovascular diseases, rheumatoid arthritis, Alzheimer's disease, lung fibrosis and inflammatory bowel disease, as well as periodontal disease [95–99]. These essential fatty acids and their precursor, the vegetable oil α-linoleic acid, are widely believed to act by prevention of conversion of arachidonate to proinflammatory eicosanoids or serving as an alternative substrate producing less-potent products. Novel oxygenated products of ω–3 PUFAs generated by enzymatic pathways similar to the production of LX are called 'resolvins', and exhibit potent actions in the resolution of inflammation. Similar in action to LX, resolvins are endogenous mediators produced from exogenous or dietary fatty acids [70, 100]. The specific members of the resolvin family (resolvin E and D series) have potent biological actions demonstrable in the nanomolar to picomolar range in vitro and in vivo [64, 100–104]. The receptors for resolvins or the E series are the chemR23 receptor expressed on macrophages and BLT1 on neutrophils [105], which are distinct from the LX receptor (FPRL1). These pathways account for, in large part, the systemic anti-inflammatory activity of ω–3 PUFA in the diet [95, 106–108]. Similar to LX, when aspirin is introduced, ω–3 PUFA are metabolized into new resolving molecules that are more stable and therefore longer acting [100] (fig. 3).

Paradigm Shift in the Pharmacologic Management of Periodontitis

Homeostasis is a fundamental characteristic of living things. It is the maintenance of the internal environment within tolerable limits, and is often described as a process of balance. Physiologic resolution of a well-orchestrated inflammatory response is essential to maintain homeostasis at the cellular and tissue levels [109]. Acute local inflammation in healthy individuals is self-limiting, and resolves by means of an active series of termination events [110]. When an inflammatory event is terminated, the complete clearance of invading leukocytes from the inflammation site is required. Complete return to homeostasis cannot occur while neutrophils remain in the lesion. If the process is incomplete, there is destruction of the cellular matrix and scarring and fibrosis can occur, resulting in incomplete healing or persistent (chronic) inflammation [34]. In many inflammatory conditions, such as periodontitis, incomplete neutrophil clearance can impact the conversion of an acute gingivitis to chronic periodontitis [18]. Scarring or fibrosis during periodontal healing prevents the return to homeostasis, which causes failure of regeneration of the periodontal tissues.

The potential of pro-resolving molecules such as RvE1 for the treatment of established periodontitis was explored using a rabbit model (fig. 4). In the RvE1-treated group, inflammation was completely eliminated; pocket depth was returned to normal and soft tissues returned to healthy levels and appearance (fig. 5). Restoration of crestal bone height, elimination of infrabony defects, and regeneration of new

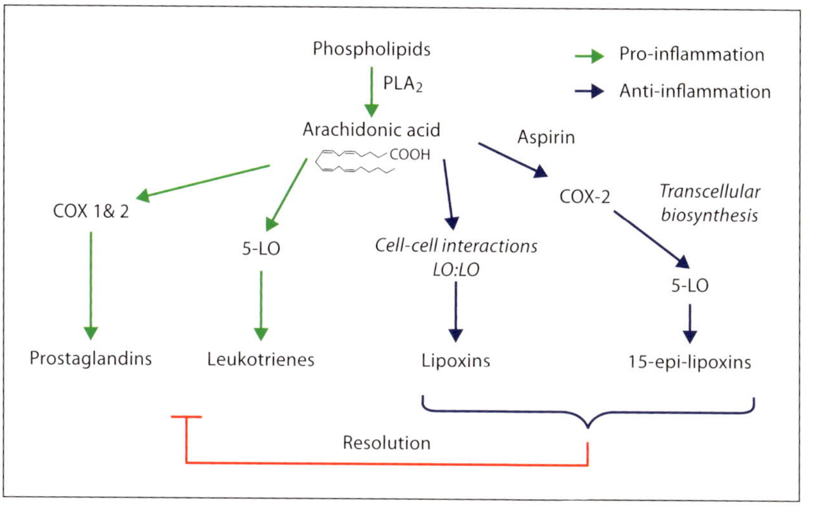

Fig. 3. Endogenous lipid mediators of inflammation. Arachidonate, an ω–6 fatty acid, is cleaved from the cell membrane by PLA_2 and becomes the substrate for two major enzyme groups; the cyclooxygenases (COX-1 and COX-2) and the lipoxygenases (5-LO, 12-LO and 15-LO). The products of these enzymes are potent proinflammatory mediators. Late in the inflammatory response, when the cells containing the LO and the LO products are in high concentration, there is a crossover of LO activity through a process called 'transcellular biosynthesis' or cell:cell (LO: LO) interactions. The resulting products are arachidonate molecules that are substituted at two sites rather than one. The enzymatic products of new double-substituted molecules are termed 'lipoxins' and represent a new class of molecule with distinct receptors and pleomorphic functions depending on the cell type driving the resolution of inflammation.

cementum, connective tissue and bone with an organized periodontal ligament were signs of complete regeneration of tissues to pre-disease levels [111]. Periodontitis was also shown to have a systemic impact elevating the levels of IL-1β and CRP in all animals. Oral topical RvE1 therapy reduced systemic IL-1β and CRP levels. Rabbits treated with RvE1 showed an essentially complete recovery without any signs of local and systemic inflammation [111] suggesting a complete return to tissue homeostasis.

Conclusion

Inflammation is a protective host response to infectious challenge and tissue injury that, if unregulated, can lead to loss of tissue structure as well as function. Inflammation is a self-limiting and self-resolving system with endogenous regulators and counter-regulatory mechanisms. Failure of regulation of inflammation leads to uncontrolled responses and tissue damage. Restoration or regeneration of tissues lost to disease is problematic in many human diseases, such as arthritis and periodontitis. Therapies that are aimed at regeneration of lost tissues attempt to

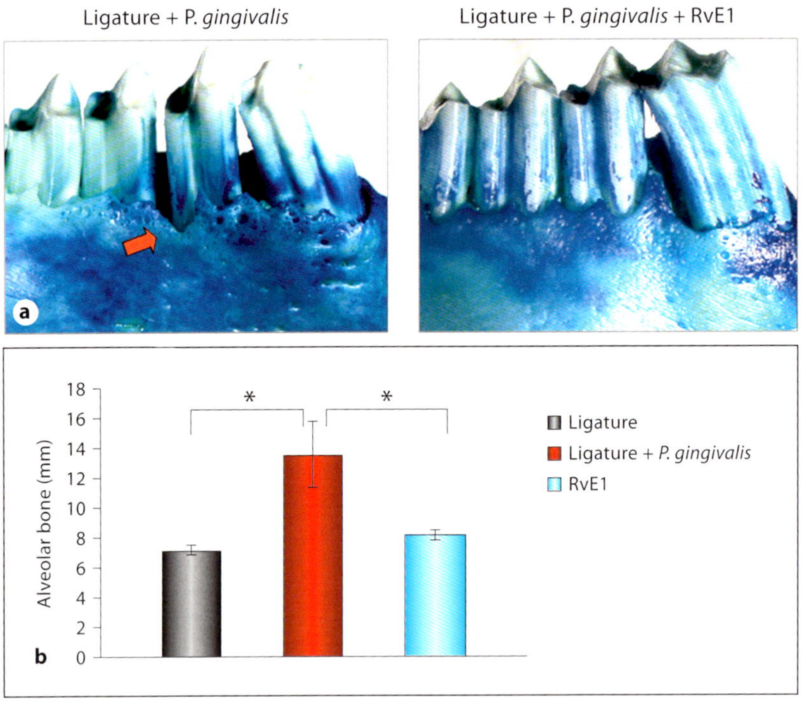

Fig. 4. Prevention of periodontitis. In a rabbit ligature model of periodontitis induced by the human pathogen *P. gingivalis*, RvE1 was evaluated for prevention of periodontitis over 6 weeks. **a** Topical application of 4 μg of RvE1 three times per week prevented periodontal inflammation and bone loss, while in the disease group (ligature + *P. gingivalis*), profound inflammation and obvious bone loss was observed (arrow). **b** Direct measurement of alveolar bone loss revealed that RvE1 treatment significantly inhibited bone loss. * $p < 0.05$.

recapitulate development. Hence, tissue engineering strategies add back growth factors, substrates and other molecules to mimic development. In the context of periodontitis, the discovery and development of pro-resolving agents emphasizes the role of local inflammation in tissue regeneration. Endogenous anti-inflammatory and pro-resolving compounds and pathways are important in resolution, catabasis, and the metabolic, cellular and tissue events leading to a return to homeostasis, but also have therapeutic potential for the treatment of inflammatory diseases. The potential use of LX, resolvins and protectins opens new avenues for the design of resolution-targeted therapies to control unwanted side effects of aberrant inflammation.

The pharmacologic use of potent 'resolution agonists' to orchestrate the return of the tissue to homeostasis and rescue resolution deficits has a net advantage over the traditional anti-inflammatory by controlling excessive cellular activity, but also enhancing clearance of bacteria and dead cells. Available data and the results of ongoing studies demonstrate that natural pathways of resolution can be used to limit

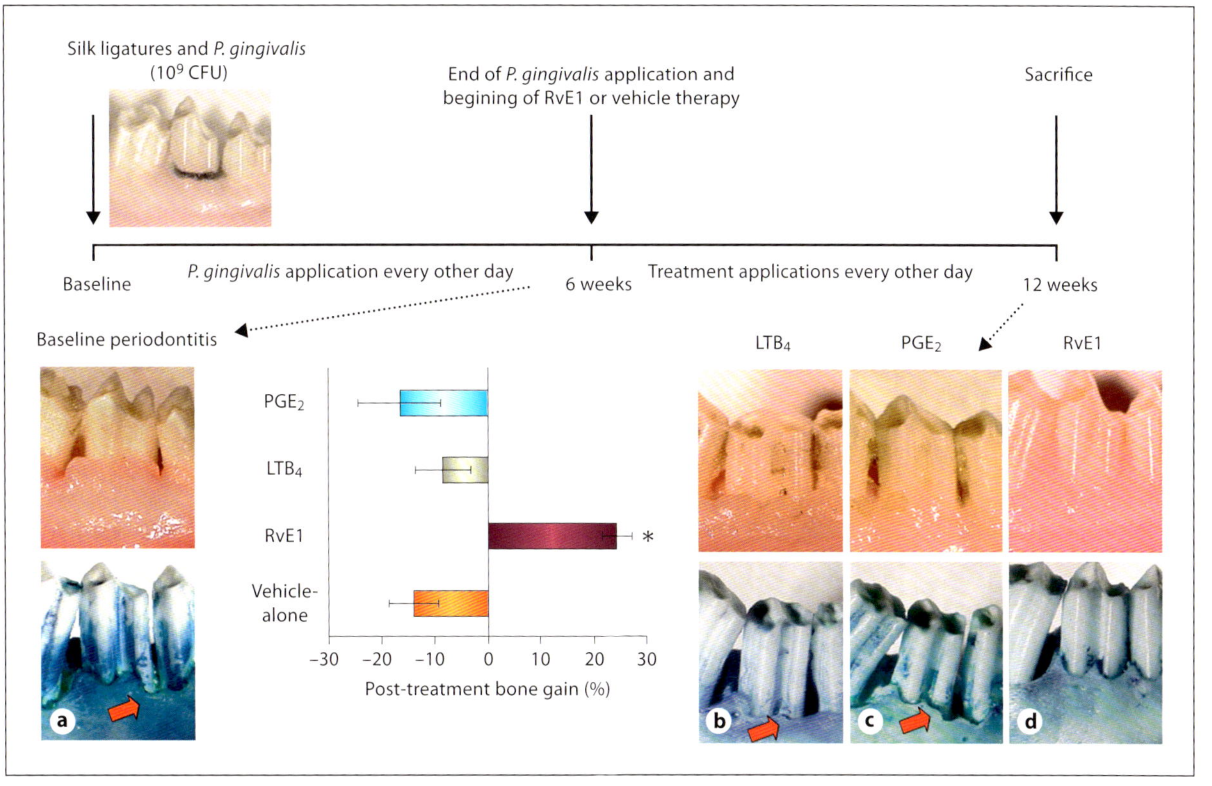

Fig. 5. RvE1 monotherapy for periodontitis. Periodontitis was induced in New Zealand white rabbits with 3–0 silk ligature and topical application of the human periodontal pathogen *P. gingivalis* for 6 weeks (baseline periodontitis). Established disease was treated for 6 weeks with either RvE1, PGE_2, LTB_4 or ethanol (vehicle). **a** Periodontal disease was obvious with all characteristics of human periodontitis including soft and hard tissue destruction at 6 weeks. **b**, **c** At 12 weeks, the periodontal disease dramatically progressed in the LTB_4- and PGE_2-treated groups. Bone destruction is clearly seen on the defleshed specimens (arrows). **d** Following 6 weeks of treatment, topical application of 4 μg RvE1 three times per week resulted in restoration of tissues to healthy levels. The bar graph shows alveolar bone loss for all animals directly measured on defleshed jaws. RvE1 treatment resulted in significant bone and soft tissue attachment gain, whereas other treatment groups showed progression characterized by soft tissue and bone destruction. * $p < 0.05$.

inflammation and promote healing and regeneration with a minimal risk of side effects [86, 100, 106, 112].

Periodontal disease is an important disease, not only because it is one of the most common diseases of man, but because within it, various aspects inflammatory disease progression can easily be studied. Studies that elucidate the pathogenesis of periodontal disease will add to our understanding of inflammatory disease in general. Within this context, lipid mediators not only present an exciting area of research, but also have potential for the development of novel treatment strategies. Understanding the role of pro-resolving molecules in regulating inflammatory processes and responses will also provide new insights into our understanding of the link between local and systemic diseases. Since the integrated response of the host is essential to health and disease, it is important to achieve a more complete understanding of the molecular and cellular events governing the formation and actions of endogenous mediators of resolution that appear to control the magnitude and duration of inflammation.

Acknowledgements

This work is supported in part by USPHS DE15566, DE19938 and DE16191.

References

1 Socransky SS, Haffajee AD: Periodontal microbial ecology. Periodontol 2000 2005;38:135–187.

2 Haffajee AD, Socransky SS: Microbiology of periodontal diseases: introduction. Periodontol 2000 2005;38:9–12.

3 Keyes PH, Jordan HV: Periodontal lesions in the Syrian hamster. III. Findings related to an infectious and transmissible component. Arch Oral Biol 1964; 9:377–400.

4 Newman MG, Socransky SS, Savitt ED, Propas DA, Crawford A: Studies of the microbiology of periodontosis. J Periodontol 1976;47:373–379.

5 Miller WD: The Micro-Organisms of the Human Mouth. Philadelphia, S.S. White, 1890.

6 Zambon JJ: Periodontal diseases: microbial factors. Ann Periodontol 1996;1:879–925.

7 Jorgensen MG, Slots J: Responsible use of antimicrobials in periodontics. J Calif Dent Assoc 2000; 28:185–193.

8 Hallmon WW, Rees TD: Local anti-infective therapy: mechanical and physical approaches. A systematic review. Ann Periodontol 2003;8:99–114.

9 Loesche WJ: Chemotherapy of dental plaque infections. Oral Sci Rev 1976;9:65–107.

10 Harper DS, Robinson PJ: Correlation of histometric, microbial, and clinical indicators of periodontal disease status before and after root planing. J Clin Periodontol 1987;14:190–196.

11 Magnusson I, Lindhe J, Yoneyama T, Liljenberg B: Recolonization of a subgingival microbiota following scaling in deep pockets. J Clin Periodontol 1984; 11:193–207.

12 Paster BJ, Boches SK, Galvin JL, Ericson RE, Lau CN, Levanos VA, Sahasrabudhe A, Dewhirst FE: Bacterial diversity in human subgingival plaque. J Bacteriol 2001;183:3770–3783.

13 Marsh PD, Moter A, Devine DA: Dental plaque biofilms: communities, conflict and control. Periodontol 2000 2011;55:16–35.

14 Armitage GC, Robertson PB: The biology, prevention, diagnosis and treatment of periodontal diseases: scientific advances in the United States. J Am Dent Assoc 2009;140(suppl 1):36S–43S.

15 Van Dyke TE, Serhan CN: Resolution of inflammation: a new paradigm for the pathogenesis of periodontal diseases. J Dent Res 2003;82:82–90.

16 Kinane DF, Lappin DF: Immune processes in periodontal disease: a review. Ann Periodontol 2002;7: 62–71.

17 Beertsen W, McCulloch CA, Sodek J: The periodontal ligament: a unique, multifunctional connective tissue. Periodontol 2000 1997;13:20–40.

18 Kornman KS, Page RC, Tonetti MS: The host response to the microbial challenge in periodontitis: assembling the players. Periodontol 2000 1997;14: 33–53.

19 Kinane DF, Lowe GD: How periodontal disease may contribute to cardiovascular disease. Periodontol 2000 2000;23:121–126.

20 Pasteur L: On the germ theory. Science 1881;2:420–422.

21 King LS: Dr. Koch's postulates. J Hist Med Allied Sci 1952;7:350–361.

22 Abraham EP: Discovery and development of penicillin. Dent Rec (London) 1946;66:197–200.

23 Harrison JW, Svec TA: The beginning of the end of the antibiotic era? II. Proposed solutions to antibiotic abuse. Quintessence Int 1998;29:223–229.

24 Fischman SL: The history of oral hygiene products: how far have we come in 6000 years? Periodontol 2000 1997;15:7–14.

25 Mandel ID: Why pick on teeth? J Am Dent Assoc 1990;121:129–132.

26 Loe H, Theilade E, Jensen SB: Experimental gingivitis in man. J Periodontol 1965;36:177–187.
27 Socransky SS: Relationship of bacteria to the etiology of periodontal disease. J Dent Res 1970;49:203–222.
28 Lindhe J, Nyman S: The effect of plaque control and surgical pocket elimination on the establishment and maintenance of periodontal health: a longitudinal study of periodontal therapy in cases of advanced disease. J Clin Periodontol 1975;2:67–79.
29 Krayer JW, Leite RS, Kirkwood KL: Non-surgical chemotherapeutic treatment strategies for the management of periodontal diseases. Dent Clin North Am 2010;54:13–33.
30 Ford PJ, Raphael SL, Cullinan MP, Jenkins AJ, West MJ, Seymour GJ: Why should a doctor be interested in oral disease? Expert Rev Cardiovasc Ther 2010;8:1483–1493.
31 Griffin SO, Barker LK, Griffin PM, Cleveland JL, Kohn W: Oral health needs among adults in the United States with chronic diseases. J Am Dent Assoc 2009;140:1266–1274.
32 Hajishengallis G: Too old to fight? Aging and its toll on innate immunity. Mol Oral Microbiol 2010;25:25–37.
33 Kao RT, Lee S, Harpenau L: Clinical challenges in diagnosing and monitoring periodontal inflammation. J Calif Dent Assoc 2010;38:263–270.
34 Cotran RS, Kumar V, Collins T: Robbins Pathologic Basis of Disease. Philadelphia, Saunders, 1999.
35 Kumar V, Abbas AK, Fausto N: Robbins and Cotran Pathologic Basis of Disease, ed 7. Philadelphia, Elsevier Saunders, 2005.
36 Wada K, Qiu FH, Stahl GL, Serhan CN: Inosine monophosphate and aspirin-triggered 15-epi-lipoxin A4 act in concert to regulate neutrophil trafficking: additive actions of two new endogenous anti-inflammatory mediators. J Hematother Stem Cell Res 2001;10:75–79.
37 Serhan CN, Oliw E: Unorthodox routes to prostanoid formation: new twists in cyclooxygenase-initiated pathways. J Clin Invest 2001;107: 1481–1489.
38 Van Dyke TE, Zinney W, Winkel K, Taufiq A, Offenbacher S, Arnold RR: Neutrophil function in localized juvenile periodontitis: phagocytosis, superoxide production and specific granule release. J Periodontol 1986;57:703–708.
39 Dongari-Bagtzoglou AI, Ebersole JL: Production of inflammatory mediators and cytokines by human gingival fibroblasts following bacterial challenge. J Periodontal Res 1996;31:90–98.
40 Genco RJ: Host responses in periodontal diseases: current concepts. J Periodontol 1992;63(suppl):338–355.
41 Page RC, Schroeder HE: Pathogenesis of inflammatory periodontal disease: a summary of current work. Lab Invest 1976;34:235–249.
42 Kantarci A, Oyaizu K, Van Dyke TE: Neutrophil-mediated tissue injury in periodontal disease pathogenesis: findings from localized aggressive periodontitis. J Periodontol 2003;74:66–75.
43 Kantarci A, Hasturk H, Van Dyke TE: Host-mediated resolution of inflammation in periodontal diseases. Periodontol 2000 2006;40:144–163.
44 Pouliot M, Clish CB, Petasis NA, Van Dyke TE, Serhan CN: Lipoxin A(4) analogues inhibit leukocyte recruitment to *Porphyromonas gingivalis*: a role for cyclooxygenase-2 and lipoxins in periodontal disease. Biochemistry 2000;39:4761–4768.
45 Offenbacher S, Odle BM, Gray RC, Van Dyke TE: Crevicular fluid prostaglandin E levels as a measure of the periodontal disease status of adult and juvenile periodontitis patients. J Periodontal Res 1984;19:1–13.
46 Kelley DS: Modulation of human immune and inflammatory responses by dietary fatty acids. Nutrition 2001;17:669–673.
47 Benito P, Nelson GJ, Kelley DS, Bartolini G, Schmidt PC, Simon V: The effect of conjugated linoleic acid on platelet function, platelet fatty acid composition, and blood coagulation in humans. Lipids 2001;36:221–227.
48 O'Banion MK, Winn VD, Young DA: cDNA cloning and functional activity of a glucocorticoid-regulated inflammatory cyclooxygenase. Proc Natl Acad Sci U S A 1992;89:4888–4892.
49 O'Banion MK: Cyclooxygenase-2:molecular biology, pharmacology, and neurobiology. Crit Rev Neurobiol 1999;13:45–82.
50 Samuelsson B: The leukotrienes: an introduction. Adv Prostaglandin Thromboxane Leukot Res 1982;9:1–17.
51 Samuelsson B: An elucidation of the arachidonic acid cascade: discovery of prostaglandins, thromboxane and leukotrienes. Drugs 1987;33(suppl 1):2–9.
52 Serhan CN, Drazen JM: Antiinflammatory potential of lipoxygenase-derived eicosanoids: a molecular switch at 5 and 15 positions? J Clin Invest 1997;99:1147–1148.
53 Funk CD: Prostaglandins and leukotrienes: advances in eicosanoid biology. Science 2001;294:1871–1875.
54 Bandeira-Melo C, Bozza PT, Weller PF: The cellular biology of eosinophil eicosanoid formation and function. J Allergy Clin Immunol 2002;109:393–400.
55 Vane J, Botting R: Inflammation and the mechanism of action of anti-inflammatory drugs. Faseb J 1987;1:89–96.

56 Vane JR, Botting RM: Mechanism of action of anti-inflammatory drugs. Adv Exp Med Biol 1997;433: 131–138.
57 Vane JR, Botting RM: Mechanism of action of non-steroidal anti-inflammatory drugs. Am J Med 1998;104:2S–8S, discussion 21S–22S.
58 Flower RJ: Studies on the mechanism of action of anti-inflammatory drugs: a paper in honour of John Vane. Thromb Res 2003;110:259–263.
59 Samuelsson B, Dahlen SE, Lindgren JA, Rouzer CA, Serhan CN: Leukotrienes and lipoxins: structures, biosynthesis, and biological effects. Science 1987; 237:1171–1176.
60 Drazen JM, Israel E, O'Byrne PM: Treatment of asthma with drugs modifying the leukotriene pathway. N Engl J Med 1999;340:197–206.
61 Jackson LM, Hawkey CJ: COX-2 selective non-steroidal anti-Inflammatory drugs: do they really offer any advantages? Drugs 2000;59:1207–1216.
62 Mamdani M, Rochon P, Juurlink DN, Anderson GM, Kopp A, Naglie G, Austin PC, Laupacis A: Effect of selective cyclooxygenase 2 inhibitors and naproxen on short-term risk of acute myocardial infarction in the elderly. Arch Intern Med 2003;163: 481–486.
63 Mengle-Gaw LJ, Schwartz BD: Cyclooxygenase-2 inhibitors: promise or peril? Mediators Inflamm 2002;11:275–286.
64 Serhan CN, Clish CB, Brannon J, Colgan SP, Chiang N, Gronert K: Novel functional sets of lipid-derived mediators with antiinflammatory actions generated from omega-3 fatty acids via cyclooxygenase 2-nonsteroidal antiinflammatory drugs and transcellular processing. J Exp Med 2000;192:1197–1204.
65 Serhan CN: Lipoxins and aspirin-triggered 15-epi-lipoxin biosynthesis: an update and role in anti-inflammation and pro-resolution. Prostaglandins Other Lipid Mediat 2002;68–69: 433–455.
66 Serhan CN, Fiore S, Levy BD: Cell-cell interactions in lipoxin generation and characterization of lipoxin A4 receptors. Ann N Y Acad Sci 1994;744:166–180.
67 Badr KF, DeBoer DK, Schwartzberg M, Serhan CN: Lipoxin A4 antagonizes cellular and in vivo actions of leukotriene D4 in rat glomerular mesangial cells: evidence for competition at a common receptor. Proc Natl Acad Sci U S A 1989;86:3438–3442.
68 Lee TH, Lympany P, Crea AE, Spur BW: Inhibition of leukotriene B4-induced neutrophil migration by lipoxin A4:structure-function relationships. Biochem Biophys Res Commun 1991;180:1416–1421.
69 Serhan CN, Romano M: Lipoxin biosynthesis and actions: role of the human platelet LX-synthase. J Lipid Mediat Cell Signal 1995;12:293–306.
70 Serhan CN, Clish CB: Aspirin triggered lipid mediators: novel inhibitors of leucocyte trafficking. Thorax 2000;55(suppl 2):S10–S12.
71 Serhan CN: A search for endogenous mechanisms of anti-inflammation uncovers novel chemical mediators: missing links to resolution. Histochem Cell Biol 2004;122:305–321.
72 Claria J, Serhan CN: Aspirin triggers previously undescribed bioactive eicosanoids by human endothelial cell-leukocyte interactions. Proc Natl Acad Sci U S A 1995;92:9475–9479.
73 Birnbaum Y, Ye Y, Lin Y, Freeberg SY, Huang MH, Perez-Polo JR, Uretsky BF: Aspirin augments 15-epi-lipoxin A4 production by lipopolysaccharide, but blocks the pioglitazone and atorvastatin induction of 15-epi-lipoxin A4 in the rat heart. Prostaglandins Other Lipid Mediat 2007;83:89–98.
74 Fiore S, Serhan CN: Lipoxin A4 receptor activation is distinct from that of the formyl peptide receptor in myeloid cells: inhibition of CD11/18 expression by lipoxin A4-lipoxin A4 receptor interaction. Biochemistry 1995;34:16678–16686.
75 Serhan CN, Sheppard KA: Lipoxin formation during human neutrophil-platelet interactions: evidence for the transformation of leukotriene A4 by platelet 12-lipoxygenase in vitro. J Clin Invest 1990; 85:772–780.
76 Serhan CN, Maddox JF, Petasis NA, Akritopoulou-Zanze I, Papayianni A, Brady HR, Colgan SP, Madara JL: Design of lipoxin A4 stable analogs that block transmigration and adhesion of human neutrophils. Biochemistry 1995;34:14609–14615.
77 Levy BD, Drazen JM, Serhan CN: Agonist-induced lipoxin A4 generation in vitro and in aspirin-sensitive asthmatics: detection by a novel lipoxin A4-ELISA. Adv Exp Med Biol 1997;400B:611–615.
78 Reddy BS, Rao CV, Seibert K: Evaluation of cyclooxygenase-2 inhibitor for potential chemopreventive properties in colon carcinogenesis. Cancer Res 1996; 56:4566–4569.
79 DuBois RN, Giardiello FM, Smalley WE: Nonsteroidal anti-inflammatory drugs, eicosanoids, and colorectal cancer prevention. Gastroenterol Clin North Am 1996;25:773–791.
80 Brady HR, Serhan CN: Lipoxins: putative braking signals in host defense, inflammation and hypersensitivity. Curr Opin Nephrol Hypertens 1996;5: 20–27.
81 Serhan CN: Eicosanoids in leukocyte function. Curr Opin Hematol 1994;1:69–77.
82 Serhan CN: Lipoxin biosynthesis and its impact in inflammatory and vascular events. Biochim Biophys Acta 1994;1212:1–25.

83 Serhan CN, Fiore S: Lipoxin recognition sites of human neutrophils. Adv Prostaglandin Thromboxane Leukot Res 1994;22:317–326.

84 Nathan C: Points of control in inflammation. Nature 2002;420:846–852.

85 Levy BD, Clish CB, Schmidt B, Gronert K, Serhan CN: Lipid mediator class switching during acute inflammation: signals in resolution. Nat Immunol 2001;2:612–619.

86 Levy BD, Lukacs NW, Berlin AA, Schmidt B, Guilford WJ, Serhan CN, Parkinson JF: Lipoxin A4 stable analogs reduce allergic airway responses via mechanisms distinct from CysLT1 receptor antagonism. Faseb J 2007;21:3877–3884.

87 Bannenberg G, Moussignac RL, Gronert K, Devchand PR, Schmidt BA, Guilford WJ, Bauman JG, Subramanyam B, Perez HD, Parkinson JF, Serhan CN: Lipoxins and novel 15-epi-lipoxin analogs display potent anti-inflammatory actions after oral administration. Br J Pharmacol 2004;143:43–52.

88 O'Meara YM, Brady HR: Lipoxins, leukocyte recruitment and the resolution phase of acute glomerulonephritis. Kidney Int Suppl 1997;58:S56–S61.

89 Ye XH, Wu Y, Guo PP, Wang J, Yuan SY, Shang Y, Yao SL: Lipoxin A4 analogue protects brain and reduces inflammation in a rat model of focal cerebral ischemia reperfusion. Brain Res 2010;1323:174–183.

90 Papayianni A, Serhan CN, Phillips ML, Rennke HG, Brady HR: Transcellular biosynthesis of lipoxin A4 during adhesion of platelets and neutrophils in experimental immune complex glomerulonephritis. Kidney Int 1995;47:1295–1302.

91 Brezinski DA, Nesto RW, Serhan CN: Angioplasty triggers intracoronary leukotrienes and lipoxin A4: impact of aspirin therapy. Circulation 1992;86:56–63.

92 Stenke L, Reizenstein P, Lindgren JA: Leukotrienes and lipoxins – new potential performers in the regulation of human myelopoiesis. Leuk Res 1994;18:727–732.

93 Sanak M, Levy BD, Clish CB, Chiang N, Gronert K, Mastalerz L, Serhan CN, Szczeklik A: Aspirin-tolerant asthmatics generate more lipoxins than aspirin-intolerant asthmatics. Eur Respir J 2000;16:44–49.

94 Serhan CN, Jain A, Marleau S, Clish C, Kantarci A, Behbehani B, Colgan SP, Stahl GL, Merched A, Petasis NA, Chan L, Van Dyke TE: Reduced inflammation and tissue damage in transgenic rabbits overexpressing 15-lipoxygenase and endogenous anti-inflammatory lipid mediators. J Immunol 2003;171:6856–6865.

95 Calder PC: *n*–3 Fatty acids and cardiovascular disease: evidence explained and mechanisms explored. Clin Sci (Lond) 2004;107:1–11.

96 Rahman MM, Bhattacharya A, Fernandes G: Docosahexaenoic acid is more potent inhibitor of osteoclast differentiation in RAW 264.7 cells than eicosapentaenoic acid. J Cell Physiol 2008;214:201–209.

97 Eickmeier O, Hilberath JN, Zielen S, Haworth O: Pro-resolving lipid mediators in inflammatory lung diseases. Pneumologie 2011;65:149–158.

98 Calder PC: Polyunsaturated fatty acids, inflammatory processes and inflammatory bowel diseases. Mol Nutr Food Res 2008;52:885–897.

99 Calder PC, Zurier RB: Polyunsaturated fatty acids and rheumatoid arthritis. Curr Opin Clin Nutr Metab Care 2001;4:115–121.

100 Serhan CN, Hong S, Gronert K, Colgan SP, Devchand PR, Mirick G, Moussignac RL: Resolvins: a family of bioactive products of omega-3 fatty acid transformation circuits initiated by aspirin treatment that counter proinflammation signals. J Exp Med 2002;196:1025–1037.

101 Lukiw WJ, Cui JG, Marcheselli VL, Bodker M, Botkjaer A, Gotlinger K, Serhan CN, Bazan NG: A role for docosahexaenoic acid-derived neuroprotectin D1 in neural cell survival and Alzheimer disease. J Clin Invest 2005;115:2774–2783.

102 Hong S, Gronert K, Devchand PR, Moussignac RL, Serhan CN: Novel docosatrienes and 17S-resolvins generated from docosahexaenoic acid in murine brain, human blood, and glial cells: autacoids in anti-inflammation. J Biol Chem 2003;278:14677–14687.

103 Marcheselli VL, Hong S, Lukiw WJ, Tian XH, Gronert K, Musto A, Hardy M, Gimenez JM, Chiang N, Serhan CN, Bazan NG: Novel docosanoids inhibit brain ischemia-reperfusion-mediated leukocyte infiltration and pro-inflammatory gene expression. J Biol Chem 2003;278:43807–43817.

104 Mukherjee PK, Marcheselli VL, Serhan CN, Bazan NG: Neuroprotectin D1: a docosahexaenoic acid-derived docosatriene protects human retinal pigment epithelial cells from oxidative stress. Proc Natl Acad Sci U S A 2004;101:8491–8496.

105 Arita M, Ohira T, Sun YP, Elangovan S, Chiang N, Serhan CN: Resolvin E1 selectively interacts with leukotriene B4 receptor BLT1 and ChemR23 to regulate inflammation. J Immunol 2007;178:3912–3917.

106 Hasturk H, Kantarci A, Ohira T, Arita M, Ebrahimi N, Chiang N, Petasis NA, Levy BD, Serhan CN, Van Dyke TE: RvE1 protects from local inflammation and osteoclast-mediated bone destruction in periodontitis. Faseb J 2006;20:401–403.

107 Hudert CA, Weylandt KH, Lu Y, Wang J, Hong S, Dignass A, Serhan CN, Kang JX: Transgenic mice rich in endogenous omega-3 fatty acids are protected from colitis. Proc Natl Acad Sci U S A 2006;103:11276–11281.

108 Schwab JM, Chiang N, Arita M, Serhan CN: Resolvin E1 and protectin D1 activate inflammation-resolution programmes. Nature 2007;447:869–874.

109 Serhan CN, Savill J: Resolution of inflammation: the beginning programs the end. Nat Immunol 2005;6:1191–1197.

110 Serhan CN: Novel chemical mediators in the resolution of inflammation: resolvins and protectins. Anesthesiol Clin 2006;24:341–364.

111 Hasturk H, Kantarci A, Goguet-Surmenian E, Blackwood A, Andry C, Petasis N, Serhan C, Van Dyke T: Resolvin E1 Regulates Inflammation at the cellular and tissue level and restores tissue homeostasis in vivo. J Immunol 2007;179:7021–7029.

112 Arita M, Clish CB, Serhan CN: The contributions of aspirin and microbial oxygenase to the biosynthesis of anti-inflammatory resolvins: novel oxygenase products from omega-3 polyunsaturated fatty acids. Biochem Biophys Res Commun 2005;338:149–157.

Hatice Hasturk
The Forsyth Institute, Suite: 1757
245 First Street
Cambridge, MA 02142 (USA)
Tel. +1 617 892 8499, E-Mail hhasturk@forsyth.org

Author Index

Subject Index